Principles of Planned Maintenance

Principles of Planned Maintenance

R. H. Clifton

CEng MIMechE
Lecturer, Mechanical and Production Engineering Department
Medway and Maidstone College of Technology

Edward Arnold

First published 1974
by Edward Arnold (Publishers) Ltd.
25 Hill Street, London W1X 8LL

ISBN: 0 7131 3317 1

Text set in 10/11 pt. Photon Baskerville, printed by photolithography, and bound in Great Britain at The Pitman Press, Bath

Preface

In recent years, the whole philosophy of plant maintenance has been gradually changing. The advent of automation, increased production levels, rigid production schedules, the high cost of capital equipment and increased machine utilization have all highlighted the fact that maintenance can no longer be considered simply as an adjunct to the production process, but must be regarded as an integral part of it. It is therefore logical that as much forethought and planning should be devoted to the maintenance aspect as to any of the other engineering activities. Many companies have already proved the value of planning and organizing their maintenance operations on a scientific basis, in close conjunction with their production requirements, while many other companies are actively considering ways and means of doing so.

A number of Examination and Training Boards have also recognized the importance of the subject by including Planned Maintenance for study in their respective syllabuses.

This book has been written with two main objectives:

(i) to assist students preparing for examinations in which Planned Maintenance forms a constituent part of the syllabus; and

(ii) to provide a practical guide for persons in industry who may be responsible for introducing, implementing and subsequently operating a planned maintenance system.

This book is not set out as a complete work of reference, nor does it provide a rigid system that can be applied universally to accommodate every situation; each maintenance system must be tailored to suit particular needs and conditions. Differing circumstances do not change the principles, only the method of their application.

An attempt has been made to express these basic principles in a simple, practical manner to enable the reader to obtain a clear understanding of the 'mechanics' of planned maintenance. The various components of a system can then be modified, assembled and adjusted to suit individual circumstances.

Maintenance technicians following the Objective Training Programme recommended by the Engineering Industry Training Board will find the book particularly applicable, and most useful when studying the Maintenance Planning module of the course.

Similarly, the Maintenance Organization and Maintenance Project content of the City and Guilds of London, Mechanical Engineering Technicians Course No. 255, Part II – Plant Maintenance and Works Services, Part III – Plant Engineering, is amply covered by the text.

Various professional bodies require their students to have a working knowledge of Planned Maintenance. In this respect it is anticipated that the material will be of special interest to persons studying for the following examinations:

Diploma of Management Studies.
British Institute of Management.
The Institution of Cost and Works Accountants.
The Institute of Work Study Practitioners
The Council of Engineering Institutions – Part 2.

The term *Planned Maintenance* has various interpretations, so too have the different terms used to describe various other maintenance procedures; the use of these ambiguous terms has led to confusion and misunderstanding. The need to regularize the terminology has largely been met by the British Standard *Glossary of general terms used in maintenance organization* BS 3811 : 1964[1]. An extract of this standard is given in Appendix 1. These terms and their respective definitions have been used when compiling the text of this book.

R. H. CLIFTON

Acknowledgements

The author is indebted to the following organizations for allowing him to reproduce material from their publications.

P. W. Allen & Co. Extracts on Optical Inspection Devices (section 4.2) including Figures 4.11 to 4.15.

BP Chemicals Ltd. Section 5.4 from Prior A. D. *Plant Maintenance Organization.*

British Petroleum Ltd. Figure 2.15.

British Productivity Council Corder's Maintenance Efficiency Index, from *Maintenance—Techniques and Outlook.*

British Standards Institution Figure 1.3 and Appendix 1 from BS 3811 : 1964 *Glossary of general terms used in maintenance organization.* Definition of method study and work measurement (section 4.1) and Figure 4.1 from BS 3138 : 1969 *Glossary of terms used in work study.* Copies of the standard are obtainable from BSI, 2 Park Street, London W1A 2BS.

Business Books Ltd. Suggested maintenance schedules for fire fighting appliances (section 6.2) from Clements R. and Parks D. Ed. (1966) *Manual of Maintenance.*

Central Electricity Generating Board Figure 2.13 from *Modern Power Station Practice,* Vol. 7, (1971).

City and Guilds of London Institute Examination questions based on those originally set for the Mechanical Engineering Technician Course, Parts II and III, (Appendix 3).

Dawe Instruments Ltd. Figures 4.5 to 4.8.

Factory Publications Ltd. Extracts in sections 5.2 and 5.3 from *Maintenance Engineering.*

HMSO (The Controller) Figure 2.20 and extracts in sections 1.6 and 2.7 from *Planned Maintenance,* issued by the Ministry of Technology (1966). Figure 2.16 from *Goods Vehicle Tester's Manual* (1972). Extracts from the Factories Acts (Appendix 2).

Institute of Marine Engineers Extracts in section 3.5 from Falconer, W. H. (1957) Some aspects of the application of planned maintenance to marine engineering. *Trans. Inst. mar. Engrs,* **69,** 2, 37.

Institution of Mechanical Engineers Extracts in Section 4.4 and Figure 4.27 from Parkes D. (1971) *The problems of the maintenance engineer, J. Instn mech. Engrs,* **18,** 5, 174.

Kalamazoo Business Systems Ltd. Figures 2.5, 2.6, 2.9, 2.12, 3.3, 5.2–5.4 and 6.12.

Mobil Oil Co Ltd. Sections 5.5 to 5.8, and Figure 3.2 from *Machine Lubrication Control* and *The Mobil Preventive Maintenance System,* Mobil Technical Bulletins.

Pitman Publishing Ltd. Figure 4.3 from Currie R. M. *Work Study.*

Production Engineering Research Association Figure 3.1.

Roneo Ltd. Figures 2.5, 2.7 and 5.12.

Arthur Taylor Ltd. Section 2.3 and Figure 2.2.

Wells–Krautkramer Ltd. Figures 4.9 and 4.10.

R. H. Clifton

Contents

1

The Objectives of Planned Maintenance

1.1. Maintenance in industry

The object of forming and operating an industrial company is to *make a profit*. No business can continue to function unless a profit is made. These facts, although obvious, are frequently overlooked by many persons who are not intimately concerned with company management.

All activities within a company should be so organized and co-ordinated that their overall effect and ultimate aim is to increase the profitability of the business. The introduction of any new activity, or the modification of an existing one, should be critically examined to ensure that it assists the company to achieve this objective.

Industry functions by producing saleable goods which can subsequently be sold for a profit to enable the company to continue to operate. The whole structure of the company is dependent upon the production and sale of these goods. Any loss or reduction of production results in a loss or reduction of profit.

Not only must industry produce saleable goods but, in order to sell them against competition, it must also ensure that the goods are:

The right price
The right quality
Produced and delivered at the right time.

To comply with these conditions, production must be carried out in the most efficient and economical manner. The plant must operate efficiently and accurately at the required level of production. There must be no unscheduled stoppages.

Intense competition together with a rapidly advancing technology have wrought many changes in the pattern and outlook of industry. New products are continually being developed, new techniques, processes, systems and methods are being applied. Automation is increasing. Production levels are being raised while rigid schedules must be adhered to. Continual efforts are made to reduce or stabilize manufacturing costs in spite of rising material and labour costs, thus making increased machine utilization an economic necessity.

These factors, although directly the concern of the production department, reflect back decisively to the maintenance department. As new production methods and machines are developed and introduced, so the outlook of the maintenance department must be progressive, employing the latest techniques to keep pace with the advancement of production technology. Previously, when machines were relatively simple and production schedules were not so rigid, maintenance presented fewer problems. However, with the emergence of complex modern machines and very tight production planning, the situation has changed considerably.

Automation, modern plant and equipment require heavy capital expenditure, making 'downtime' extremely costly. To ensure maximum plant availability and reliability, regular maintenance must be carried out. This maintenance must be carefully planned in conjunction with production requirements and schedules so that it causes the minimum stoppage and loss of production. Inadequate maintenance can lead to damage which is extremely costly not only in repairs but also in lost production.

With the increasing complexity, sophistication and cost of modern equipment the maintenance department is an indispensable part of a production system. Production departments are depending more and more upon the skills and organization of the maintenance department and it is accepted that maintenance is now a specialized function of growing importance and size.

Even though the importance of the maintenance function is steadily increasing, and is indeed a dominant factor in many industries, it must be viewed in the correct perspective to ensure that it does not obscure the real objectives and does not become a case of 'the tail wagging the dog.'

Figure 1.1 illustrates the typical organizational structure of a medium-sized engineering company. Profit is made by the company only upon the goods that are produced in its workshops. The *only* persons who are actually producing these saleable goods are those shown within the shaded area – the foundry workers, machine operators and the fitters. (In the case of chemical process plants, oil refineries and similar production units, the persons producing the saleable product would be the process operators.)

The company is not selling plant maintenance, canteen services, inspection, accountancy services, transportation or supervision. All the functions shown outside the shaded area, although necessary for the efficient operation and success of the business, do not in themselves produce any of the goods which are bought by the customers. They cannot exist simply by their own efforts but only by reason of their contribution to the overall functioning of the company.

The reason for having a maintenance department is to ensure that the machines, buildings, services and equipment are operating at the required level of productive efficiency and are available when required;

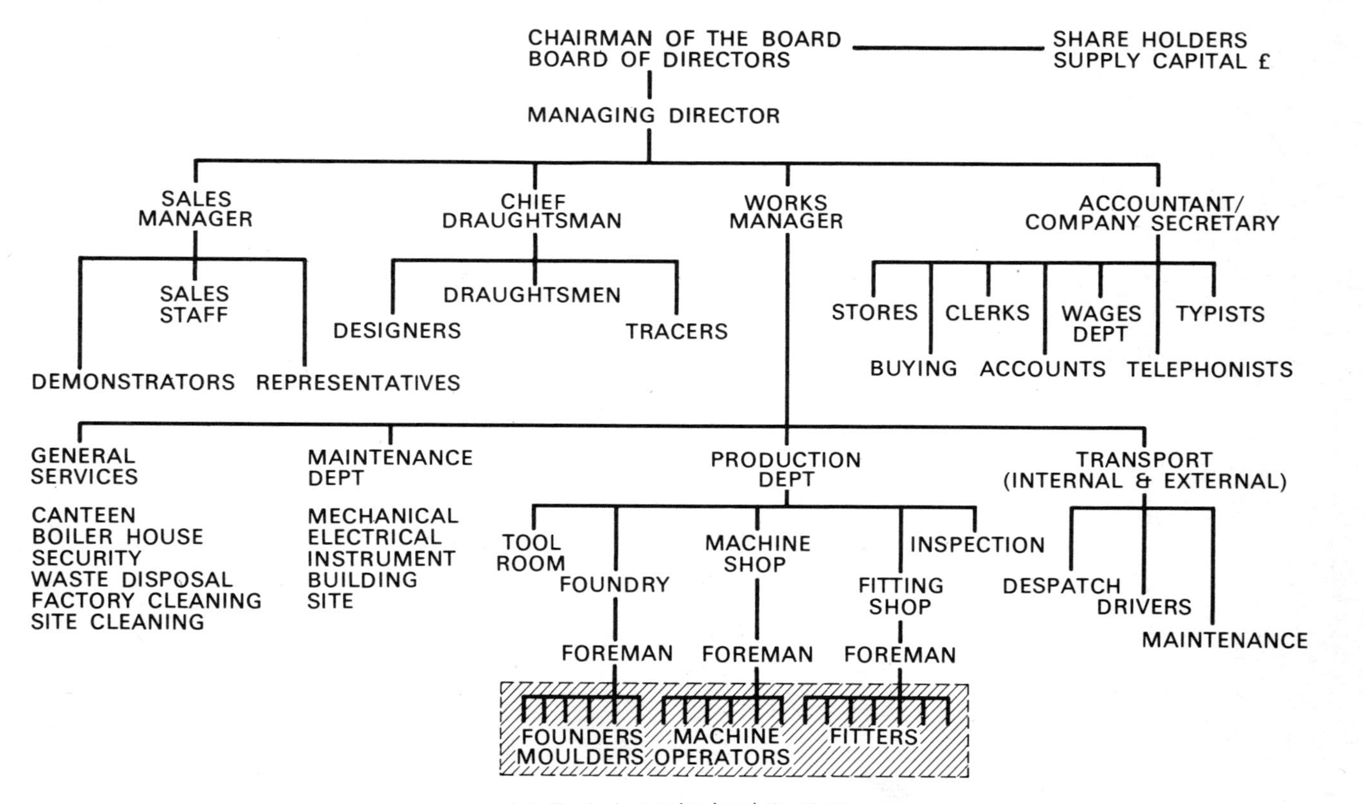

1.1 Typical organizational structure

its function is to be a *service* to the production department.

This does not imply that service departments are subservient to the production department. All the activities that take place within the framework of the company can only be effective if there is co-operation by all the various sections. This co-operation must be a two-way effort, as it is equally essential that the production department co-operates with its service departments to gain their full support in the most expedient manner.

1.2. Maintenance in public services

So far, we have considered only industrial manufacturing concerns in which a stoppage or loss in production can be measured directly in terms of loss of profit. If this loss in production is the result of breakdowns due to insufficient maintenance, then the economics of standby equipment, replacement units or extra maintenance can be calculated and the profitabilities compared. Breakdowns can jeopardize delivery dates, and their effect on sales and future orders can be forecast. In each case, the cost can be estimated and used by the management to arrive at the most effective decision.

This method can also be applied partly by certain public utilities which are required to pay their way – gas boards, water companies, public transport, electricity generating and distribution boards.

In the case of commercial offices and hotels, the customer is paying for a service which includes engineering functions (electrical distribution, heating, hot and cold water supply, ventilation, lifts, etc.). A breakdown of any of these services may not directly, or immediately, cause a financial loss but would eventually reflect back on the reputation of the establishment, which in turn could affect business.

There are also many other situations in which failure of engineering and maintenance services cannot be assessed solely in financial terms.

Hospitals function continuously; a breakdown or stoppage of certain equipment could prove fatal. In such cases, it is necessary to instal standby equipment to carry out repairs and maintenance on the run. A comprehensive maintenance system is essential to ensure the equipment is reliable at all times.

The effective disposal of refuse and sewage together with an efficient drainage system are vital to public health. Any failure could contribute to the spread of pests and disease within the community. The successful functioning of this service relies upon the continuous operation of engineering plant, with standby equipment ready to deal with flooding or other emergencies. As in hospitals, measures must be taken to ensure that the equipment is maintained to minimize the risk of breakdown.

The failure of engineering services in schools, libraries and council offices would not be disasterous, or result in a loss of profit, but it would cause inconvenience.

The public services do not exist to make a profit, their prime function is to provide services for the community. The value placed upon any particular service and the overall expenditure on it vary according to the priorities of the community. These priorities frequently change, being affected by national and local policies, public opinion, available resources and public or governmental pressures. Because of these diverse and often conflicting conditions, it is impossible to formulate an overall, tangible standard by which the efficiency or economics of any service can be estimated.

Each situation must be judged upon the conditions prevailing, the aims and objectives of the service, and the boundaries within which it must operate.

Within any service, there are sectors in which the economics and efficiencies of the various factors can be measured and compared. In such cases, these measurable factors should be isolated and standards set to enable the optimum use of the existing equipment and resources to be made.

1.3. The duties of a maintenance department

The duties of a maintenance department vary from company to company, and often from factory to factory within the same company. The scope and depth to which it executes its duties depends upon company policy, size and type of plant, type of production and skilled labour available, to name but a few of the determinant factors. The maintenance department of a small factory may consist only of a handyman who will tackle all types of maintenance to a limited extent and of an elementary nature; outside specialist contractors are called in for anything beyond this. The maintenance department of a large industrial complex could be a self-contained unit employing skilled craftsmen who collectively are capable of any maintenance, overhaul or installation work within the site.

Usually, the work which the department undertakes may be divided into four main groups:

1. Pure maintenance, defined as 'work undertaken in order to keep or restore a facility to an acceptable standard'.
   (a) Maintenance of existing plant, equipment and buildings.
   (b) Rebuilding or reconditioning old plant and equipment.
2. Installation and alteration of plant, equipment and services.
   (a) Installation of new plant and equipment.
   (b) Installation of utilities – steam, water, gas heating, ventilation, electric light and power.
   (c) Installation of special services – vacuum, industrial gas and

chemical pipelines, purified water systems, compressed air systems, fire alarms.
(d) Alteration or modification of plant, equipment and buildings.
(e) Alteration or modification of utilities and special services.

3. Operation and supervision of particular utilities and special services.
(a) Boiler house operation, steam supply, power and generation.
(b) Compressed air plant and supply lines, heating and ventilating system.

4. Miscellaneous duties delegated to the maintenance department because it is often the only department with the ability to be able to handle or supervise the work. (This is particularly applicable in factories which employ a predominance of female operatives.)
(a) Factory and site cleaning services – roads, floors, windows.
(b) Waste and effluent disposal.
(c) Salvage collection and disposal.
(d) Fire fighting services.
(e) Factory security.

Whatever duties are assigned to the maintenance department, the prime reason for its existence is defined in 1(a).

'To ensure the availability and efficiency of existing plant, equipment and buildings in a manner required by the production department.'

In terms of plant operation this means that:
(a) The plant must be available for start-up when required.
(b) The plant must not break down during production runs.
(c) The plant must operate in an efficient manner at the required level of production.
(d) The downtime for maintenance must not interfere with production schedules or runs.
(e) The downtime which may be caused by a breakdown should be a minimum.

To accomplish the above conditions:
(a) There must be complete co-operation and mutual understanding between maintenance and production departments.
(b) There must be an effective maintenance policy for planning, controlling and directing all maintenance activities.
(c) The maintenance department must be well organized, adequately staffed and the personnel sufficiently trained to carry out the work.
(d) There must be progressive efforts to reduce or eliminate breakdowns.

1.4 Plant operation

If the plant is to meet production schedules with the minimum of disruption, it is logical that shut down maintenance should take place

(i) when the type or nature of production necessitates shut down, or
(ii) when the plant is shut down because of the pattern of working hours.

Shut downs due to the nature of production will vary according to the peculiarities of the process, and in many instances can only be forecast on a short range basis. This makes long term planning extremely difficult.

In many industries and factories, the pattern of the working week is well established and little change takes place throughout the year. Plant shut downs occur regularly and can be predicted with reasonable accuracy over long periods.

The pattern of working hours depends largely upon the type of industry, type of production and level of production. A factory may work eight hours per day, five days per week with occasional overtime as the need arises, whereas a continuous process industry may operate the full week of 168 hours. The hours worked by the great majority of companies lie somewhere between these two extremes. Whatever the number of hours worked, it is possible to devise a shift work system that forms a regular pattern.

Figure 1.2 summarizes different shift work systems commonly applied in industry. It tabulates against each the weekly working hours together with the time that is available to carry out shut down maintenance without interfering with production runs.

The maintenance and operation of plant and equipment is only a means to an end – the production of saleable goods at the most economical cost.

The respective functions of maintenance and operation in a modern production system are so interrelated that it is no longer possible, or desirable, to consider each as a separate, isolated element. Efforts to reduce the cost of either one of these functions may easily incur increased costs in the other, to an extent where all savings are eventually lost. It is therefore important that both the maintenance and operation of the plant are considered and planned on a unified basis with the object of achieving the minimum overall production costs.

What then is the best way to carry out these functions to obtain the required objective? There are four alternatives:

1. Operate the equipment until it breaks down – then scrap it and buy new (*replacement instead of maintenance*).
2. Operate the equipment and then sell it before it either breaks down, or requires expensive overhaul (*planned replacement*).
3. Operate the equipment until it breaks down – then repair it (*breakdown maintenance*).

System of Working			
Detail of Shift	Total Production Hours per Week	Time Available for Shut-down Maintenance Without Interfering with Production	Remarks
Single Shift (day work)			General engineering workshops
8 hrs x 5 days	40	16 hrs/day + 2 days at W/E = 128 hrs	
10 hrs x 5 days	50	14 ,, / ,, + ,, ,, ,, = 118 hrs	
12 hrs x 5 days	60	12 ,, / ,, + ,, ,, ,, = 108 hrs	
Double Shift			Batch, and semi-continuous production
2 x 8 hrs x 5 days	80	8 hrs/day + 2 days at W/E = 88 hrs	
2 x 8 hrs x 5 days + 8 hrs Sat:	88	8 ,, / ,, + 1 day + 16 hrs at W/E = 80 hrs	
2 x 8 hrs x 6 days	96	8 ,, / ,, + 1 day at W/E = 72 hrs	
Continuous Shift Work			Continuous process plants: Chemicals, Oil refineries, Steel works, Public utilities — Gas, Water, Electricity
24 hrs x 5 days	120	2 days at W/E = 48 hrs	
24 hrs x 5½ days	132	1½ ,, ,, ,, = 36 hrs	
24 hrs x 6 days	144	1 day ,, ,, = 24 hrs	
24 hrs x 7 days	168	Nil — plant is shut down only by arrangement with production department	

1 week = 7 days x 24 hrs = 168 hrs

1.2 Shift work systems employed in industry

4. Carry out regular maintenance, planned carefully in conjunction with production requirements to prevent the failure of the equipment during production runs (*preventive maintenance*).

1.5. Forms of maintenance

1. *Replacement instead of maintenance.* A factory using small, easily-replaceable, cheap equipment may find that the cost of repair exceeds the cost of replacement. Alternatively, technological advancement causes much equipment to become obsolete rapidly; consequently, it is not designed to last – when it fails it is out of date so may as well be replaced with modern equipment.

Many items are impossible to repair, while other items are designed not to be repaired – their virtue lies in their cheapness.

2. *Planned replacement.* This policy is practised frequently with vindicable results in many branches of industry, particularly those in which the equipment operates as an individual unit – machine tools, contractors plant (mobile generators, mobile compressors, mobile cranes, etc.), small power plants and motor vehicles.

Car fleets are operated invariably on this basis and many private motorists run their own vehicles in a similar manner.

Good quality equipment is purchased, and only basic maintenance (lubrication, servicing and adjustment) is carried out to keep it in good condition. It is then sold, traded in or scrapped before it looses its operational efficiency, when breakdowns and failures can be expected, or when expensive replacements and overhauls are required.

When applied in the right circumstances, this system can reduce downtime caused by breakdowns, prevent costly overhauls and minimize maintenance. It enables a company to obtain continually the advantages of modern equipment, which is particularly important in industries where technology and production methods are rapidly changing.

This system is often applied, in part, to major modules of large machines – new or reconditioned engines, gear boxes, rolls, etc. replace the old, worn ones at predetermined intervals throughout the life of the main machine.

3. *Breakdown maintenance.* Many items of plant and equipment operate as individual units, or are separate to the actual manufacturing process, so their failure would not immediately or greatly affect the overall production process or constitute a safety hazard. The cost of preventing their failure may be more than the cost of the breakdown. In this particular set of circumstances, it is justifiable financially to allow the machine to breakdown before carrying out any maintenance – lubrication and minor servicing excepted.

A factory operating a large number of similar light machines such as sewing machines or bench tools may calculate that it is easier, cheaper and quicker, with less loss of production, to allow the machines to breakdown and then replace them with reconditioned ones held in stock for such an occurrence. The old machine is then repaired or reconditioned in the workshops and held in the stores as a future replacement.

Many general engineering workshops apply only basic maintenance to their machine tools, any further maintenance is attempted only when the machine breaks down, or its efficiency becomes so low that it can no longer be tolerated.

Often, non-productive plant can be placed on a breakdown maintenance basis without any ill effect on production.

4. *Preventive maintenance.* Where a process is continuous or highly automated – chemical plants, steel mills, oil refineries, mass production and transfer lines – the cost of lost production due to breakdown can be extremely high. The failure of a small but vital piece of equipment can arrest the whole process.

Alternatively, the failure of other types of equipment – boilers, pressure vessels, lifting gear – can be dangerous.

The aim of preventive maintenance is to effect the work of inspection, servicing and adjustment, and so prevent the failure of equipment during operation. It is the anticipation of failures and the adoption of necessary preventive action before they occur.

Seldom, in practice, are situations or solutions as clear cut or precise as those quoted above. As the type of production and circumstances differ, so must the approach to the problem. Each case must be considered on its own merits – there is no universal method of maintenance that will solve all the problems. The optimum proportions of replacement, breakdown and preventive maintenance vary considerably from industry to industry, from factory to factory and indeed from machine to machine within the same factory.

In industries and factories organizing production on a continuous shift basis, or where there is substantial co-ordination of the various production units, the proportion of preventive maintenance compared with the other forms of maintenance is necessarily high, and, whenever possible, carried out while the facility is in service. Whereas, in a department on single shift working, most of the maintenance is corrective or shut down, carried out during the frequent periods when the equipment is not operating.

The size of the facility does not influence necessarily the form of maintenance to be applied, only its magnitude.

Benefits of any form of maintenance will be completely realized only by application of careful planning and organization.

1.6. Planned maintenance

Planned maintenance is not a specific type of maintenance, but the application of maintenance tackled in a scientific manner. It is the comprehensive planning of the maintenance function.

By its definition – work organized and carried out with forethought, control and records – it includes the whole range of maintenance and can apply equally to any type – replacement, breakdown or preventive, provided that:

(a) the maintenance policy has been considered carefully,
(b) the application of the policy is planned in advance,
(c) the work is controlled and directed to conform to the pre-arranged plan, and
(d) historical and statistical records are compiled and maintained to assess the results and to provide a guide for future policy.

As stated previously, the ultimate purpose of industry is to produce goods at the most economical cost. The maintenance plan must be organized to achieve this objective. It must consider the overall situation and embrace not only the maintenance and production functions but also the economics. Initially, each department – maintenance, production and accounts – will have its own pre-conceived ideas on what constitutes the best maintenance plan; but when the objective is defined and the problem studied on a combined basis, it is possible to develop a 'unified maintenance plan' that will give the best overall results.

Where a policy of planned replacement is practised, the planning is mainly of a financial nature – ensuring money is available for the purchase of new equipment, balancing depreciation costs against tax allowances, negotiating the sale, or trade in, of old equipment at the most favourable price – but by far the greater part of planned maintenance is concerned with the organization of the engineering aspect of maintenance and its relationship to production.

This maintenance may be effected in different ways; when the facility is running, when it has stopped, before it breaks down, after it breaks down, it can be planned or it can be done as and when the need arises, without any prior planning. Figure 1.3 illustrates the relationship between these various forms, the numbers under each term refer to the definition number listed in British Standard BS 3811 : 1964 an extract of which is given in the Appendix 1. In general, these terms are self explanatory and with the aid of the definitions listed require no further clarification. However, occasionally there is a misunderstanding between the use of terms emergency maintenance (10) and breakdown maintenance (04).

The terms emergency maintenance and breakdown maintenance are frequently used in an interchangeable manner, both are synonymous with sudden, unexpected breakdowns. When these occur, all available resources are urgently utilized to restore the facility to its normal operational standard in the shortest possible time.

But within the context of the standard emergency maintenance (10), is the type of failure that has not been foreseen and to which no advanced

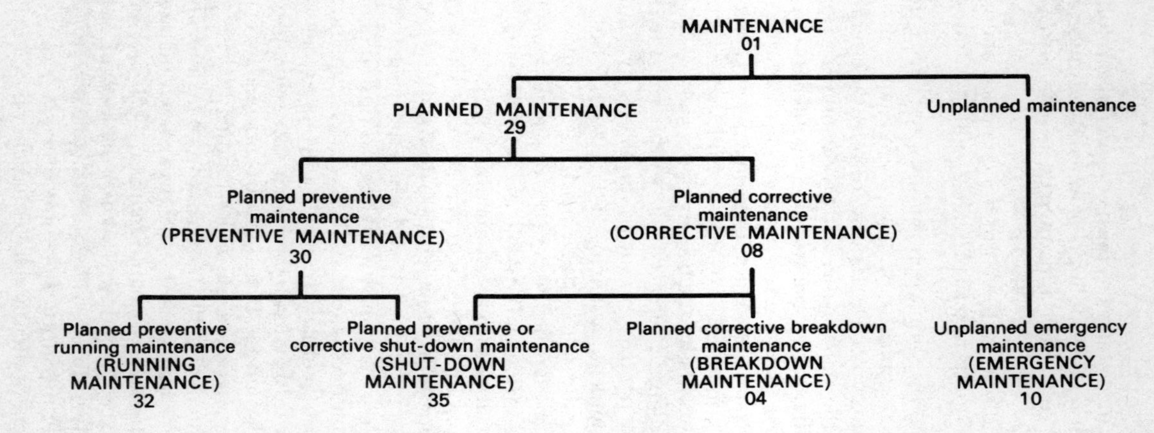

1.3 Relationships between various forms of maintenance (the figures refer to the definition number). (See Appendix 1.)

thought has been given. It is a failure that is thought could not possibly happen.

On the other hand, breakdown maintenance (04) is work carried out after a facility has failed, but its failure has been considered in advance and provision made by planning the method of repair, tools, spares and replacements.

The fundamental basis of any planned maintenance system is deciding in advance:

The individual items of plant and equipment to be maintained.
The form, method and details of how each item is to be maintained.
The tools, replacements, spares, tradesmen and time that will be required to carry out this maintenance.
The frequency at which these maintenance operations must be carried out.
The method of administering the system.
The method of analysing the results.

Expressing these basic essentials in a manner that will form the structure of a practical system; there must be:

1. A schedule of all the plant and equipment to be maintained
2. A complete schedule of all the individual tasks that must be carried out on each item of plant
3. A programme of events indicating when each task must be carried out
4. A method of ensuring that the work listed in the programme is carried out
5. A method of recording the results and assessing the effectiveness of the programme.

Even the simplest of planned maintenance schemes must be controlled, and certain facts need to be readily available to the person in control if this is to be effective. As the human memory cannot retain more than a very limited amount of information with any guarantee of accuracy, and as the scribbling of notes on odd scraps of paper – which are often lost – is equally useless and unreliable, a well organized, more permanent, system of documentation is essential.

There are many basic documentation systems available that have been designed specifically for the control and support of planned maintenance schemes, and they are all capable of being tailored to suit the individual requirements of any company or organization. Frequently, these basic systems are elaborated to provide information on costs, man hours, repair times, materials and replacements used or any other relevant data that are considered helpful. But any system that is installed should be easy to operate, involve the maintenance staff in the minimum of paper handling and recording but *must* be capable of indicating clearly:

what is to be maintained
how it is to be maintained

when it is to be maintained
is the maintenance effective.[8]

1.7. Benefits of planned maintenance

The initial organization and subsequent operation of any planned maintenance system will involve time, money and usually a considerable amount of hard work. For a company to justify this investment there must be substantial benefits. Therefore, any company considering the introduction of planned maintenance will of necessity want to know:

Can it be successfully applied in our particular case?

What benefits will result?

Planned maintenance can be applied successfully to all types of industry but its ultimate effects and benefits will differ – they depend upon the industry, the local conditions, the depth and scope, and form of application.

It is not the panacea for every maintenance problem. It will not compensate for poor workmanship, lack of tools, bad design or the maloperation of machinery by the personnel, nor will it convert worn-out obsolete equipment into modern, highly efficient units. However, the mere fact that maintenance is being considered and controlled in a systematic, constructive manner must lead to some positive benefits which are the direct result of this planning.

The essence of company planning is the optimization of its resources – men, money, materials and machinery.

Planned Maintenance can contribute to this objective as follows:

1. Greater plant availability.
   - (a) Fewer breakdowns will occur in plant that is regularly and correctly maintained.
   - (b) Maintenance is carried out when it is most convenient and will cause the minimum loss of production.
   - (c) Regular, simple maintenance results in less downtime than infrequent expensive *ad hoc* maintenance.
   - (d) Excessive length of downtime is reduced, spares and equipment demands are known in advance and are available when necessary.
2. Regular, simple servicing is cheaper than sudden expensive stop gap repairs.
3. Regular, planned servicing and adjustment maintains a continuously high level of plant output, quality, performance and efficiency.
4. Greater and more effective labour utilization.
   - (a) The volume of maintenance work is planned so that it is spread evenly throughout the year, widely fluctuating demands upon the labour force are reduced.

(b) The weekly work load is known in advance so that its allocation can be controlled effectively.
(c) The effect of a positive maintenance policy is frequently reflected in an improved personal attitude of the staff. A purposeful approach – higher morale.

5. The servicing and adjustment of equipment contained in the programme are not overlooked or omitted.

An essential part of any planned maintenance system is the collection, recording and subsequent interpretation of maintenance data. Although, primarily, these are required to assess the results and to assist forward planning, their extension offers management other benefits which increase greatly the usefulness of the system.

These other benefits usually present themselves as services to management; they are not achieved directly by planning but only record the results of it for further interpretation. Their nature will depend upon the information that is fed into, and stored by, the system.

The benefits that these services can offer may be evident in the form of:

(a) Improved budgetary control – realistic budgets can be formulated and subsequently controlled.
(b) Improved stock control of spares – realistic quantities of spares are stocked and re-ordered.
(c) Provision of information upon which management can make realistic forecasts and decisions.
(d) Focusing attention on frequently recurring jobs and types of defect, and the type, frequency and cost of individual repairs.

To provide these extra facilities, additional documentation and tabulation, over and above that normally required for basic planning, will be necessary. The nature of this paperwork will depend upon the individual circumstances and the benefits in view. Although the practical benefits may be readily recognized, the financial advantages may be less obvious, and also extremely difficult to assess until the scheme has been operating for some time.

Initially, the maintenance costs may rise as extra work is involved to catch up with the backlog, to overcome the ravages of past neglect and to bring the equipment up to the required operational standard. When this initial work is completed and the scheme is functioning normally, the costs will fall but there is no guarantee that they will fall below those prior to the commencement of the scheme. The true measure is reflected only in the ultimate overall cost of the product.

Even this ultimate cost may require careful scrutiny and adjustment as various factors affecting production costs – higher level of production, new machines, increased labour and material costs and modifications in product design and specification – may have occurred in the meantime.

2

The Basic Mechanics of Planned Maintenance Systems

What is to be maintained.
How it is to be maintained.
When it is to be maintained.
Is the maintenance effective[8]?

This chapter deals with the basic components required for the construction of a planned maintenance system. No matter how large, small, simple or sophisticated the system is, the same basic elements are common to all, though the manner in which they are used and the form they take may differ in each case, depending upon the individual circumstances.

2.1. Initiating the scheme

With the exception of very small localized arrangements, organized by an individual for his own convenience, the introduction of most planned maintenance schemes is the result of a decision taken by higher management. It is understandable that the initiative should originate at this level – not only must senior management actively promote and encourage new ideas, but it is in a position to evaluate them and has the necessary authority to see them through. Management will appreciate from past experience that the implementation and subsequent operation of any new system takes time and money, and creates certain changes within the organization. In this particular case, the maintenance and production departments will be affected most. Upper management at this stage of the proposals is not expected to formulate the exact details or operational procedures of the system, but at least it must define clearly and positively the framework within which the system must function, together with its objectives.

Planned maintenance is not different from any other human activity in which a number of persons contributes towards a common purpose inasmuch that it can be planned and carried out more efficiently if there is one person, a leader, to control and co-ordinate the mass of individual effort. In this respect, management should appoint a person to whom this particular task can be delegated. The person selected may be the chairman of a specially formed committee, the works engineer, a person already in the employ of the company, or a person specifically engaged by the company for this work. Whoever the person is, if he is to interpret the management's policy in the manner intended, he must be given a comprehensive brief detailing the policy, the boundaries within which he can operate, the results he is expected to achieve, a programme of target dates, and the resources at his disposal. He must also be given, within the terms and limits of his brief, the authority to make decisions, the power to carry them out and the responsibility for the actions he may take.

An appointment of this nature together with the proposal of planned maintenance may create initially considerable speculation, resentment and even resistance not only among particular individuals but also from whole sections of the staff. Consequently, it is most important to take any necessary preventive measures at a very early stage of the proceedings before a troublesome situation can arise and escalate. Thus, it is in the management's own interest not only to consult the staff concerned at the very beginning, informing them of the company's proposed policy and intentions, but also to keep them 'in the picture' as the scheme progresses through its various stages.

The person responsible for the scheme will be in close contact with various staff members at different levels of authority; the timing and method of his introduction as well as his own personal approach will often determine the co-operation he receives.

One of the first practical steps for any person setting up a planned maintenance organization is to make a general appraisal of the whole situation, surveying the task in hand and the available resources. A tentative approach is then worked out so that the fundamental principles can be applied in the method and manner best suited to the circumstances. This preliminary familiarization study may be thought unnecessary for a person in the employ of the company and who already has a detailed knowledge of its organization, but it is useful to reflect that the company is now proposing to change its whole philosophy of maintenance; consequently, its personnel must adapt their ideas to this new concept. With this in mind, and by taking a fresh look, old, familiar problems and situations assume an entirely new complexion. New ideas are conceived, fresh avenues of thought are opened which may eventually lead to an approach or solution not previously visualized.

2.2. The inventory

The inventory is a list of all facilities – all parts of a site, building and contents – for purposes of identification.

When the preliminaries have been completed, the first step in formulating a plan is to establish *what is to be maintained.* As most companies hold a register listing all their equipment, the required information could be extracted from this readily available source, thus saving considerable time and trouble. However, experience usually reveals that these records are not always entirely correct; additions, deletions, replacements and plant movements all occur, but occasionally fail to get recorded. Nor can layout drawings of plant arrangements be guaranteed to be accurate.

It is of little use basing a plan upon *what should be there* or *what was thought to be there.* For a plan to be workable and of any real value, it must be founded on fact – *what is actually there.* The only sure way to ascertain this is to carry out a physical inventory, listing each item and marking its position on a block plan of the area. Compiling the inventory can also provide a better understanding of the individual items and the overall production process. No matter how familiar the compiler is with the equipment and its operation, the fact that the information must be sought and considered, probably for the first time, with the object of planning the maintenance function, something will be highlighted that was not previously apparent or appreciated.

Figure 2.1 illustrates the form of an inventory sheet that would be suitable. Allowance, by way of additional columns, has been made for information which will be used at later stages of the planning. In the example shown, details contained in the Description and also in the Remarks columns were obtained from the nameplates of the respective machines, while the Identification Symbols were formulated by the method discussed in section 2.3.

Notes to assist the correct completion of the form are given below:

Identification Symbol: Each item must possess a positive means of identification (for details see section 2.3).

Description of Facility: A brief description of the item concerned; where possible, nameplate details should also be included.

Location: The department, section or area, in which the item is located, e.g. machine shop, boiler house, compressor room.

Type: When only a few items are contained in the inventory it is probably sufficient to classify them as either mechanical or electrical, but in a large industrial complex a more detailed classification is usually necessary. Methods of cost allocation, the distinct stages of a process, or specific types of plant may each be used to classify the equipment. Examples of categories which are sometimes used are listed below.

1. Major mechanical plant – pumps, compressors, boilers.
2. Major electrical plant – incoming transformers, breakers, and switchgear, rectifiers, main motors and starters.
3. Minor mechanical plant – valves, hydraulic rams.
4. Minor electrical plant – small motors and starters. Small electrical items transferable between different machines.

NVENTORY SHEET					Sheet No ....
entification ymbol	Description of Facility	Location	Type	Priority Rating	Remarks
-01-01	Centre lathe: Lang 13" swing Model J.6. Serial No 62 B/10	Maintenance work-shop	Machine tool	5	
-91-01	Electric motor (driving lathe) Brookhirst Igranic Ltd No C11360/61/2: 5 h.p.	,,	Electric (minor)	5	Type SC: 400/440 volts 3 phase: 50 cycles
-03-01	Shaper "Invicta". Type 2M B. Elliott (Machinery) Ltd. London Serial No B.E.C. 19017/4	,,	Machine tool	5	
-91-02	Electric motor (driving shaper) Brook Motors. 3 h.p. No L131 651	,,	Electric (minor)	5	A.C. class E: INT rating Frame C182: 1420 r.p.m. 400/440 V: 3 ph. 50 ~: 4·7 amp
-02-01	Milling M/C (Universal) B. Elliott (Machinery) Ltd. London Serial No B.E.C. 011236/120	,,	Machine tool	5	
-91-03	Electric motor (driving milling m/c) Newman: 3 h.p. Conn Diag No C123006 ED 30 25	,,	Electric (minor)	5	Class F: 1425 r.p.m. 400/440 V: 3 ph. 50 ~ CMR rating: 4·9 amp Frame C184 DC 1592 BB
-50-01	Steam boiler: Paxman economic No 22653: W.P. 60 P.S.I. Davy Paxman & Co Ltd. Colchester	Boiler house	Mechanical (major)	3	Space heating and water heating service
-53-01	Fuel oil burner for Paxman boiler Clyde Automatic Oil Burners – Glasgow Machine No H2 584: Inst No A8319	,,	Mechanical (major)	3	Electric supply 415 V: 3 ph. 50 ~: Fitted to above boiler
-50-02	Steam boiler: Cleaver Brooks Packaged boiler: Marshalls of Gainsborough. Model 621-300 h.p. No 96785	Boiler house	Mechanical (major)	1	Process steam Capacity 10,350 lbs/hr W.P. 150 P.S.I.: Hyd T.P. 283 Test date 17-6-68
eing installed contractors. mbol not yet ocated.	Steam boiler: Cleaver Brooks Packaged boiler: Marshalls of Gainsborough. Model 621-300 h.p. No 97522	,,	,,	1	Details as above Hyd test date 12-6-70
-60-01	Air compressor: Atlas Copco Type BT 3E: No F30 09 30 Max Press 125 P.S.I. 8·8 Kg cm² 1460 r.p.m.	Comp^r room	Mechanical (major)	1	Air supply to control instruments and gear
-94-01	Electric motor: Direct coupled to compressor. 30 h.p. Brooks Motors No X367273	,,	Electric (major)	1	Frame C180 M: 1460 r.p.m. C.M.R. rating: 39 amp 415 Δ volt: 3 ph. 50 ~:
-63-01	After-cooler (air comp^r) Finned. Tubes: air cooled, fan blown. Type TD3: No 910	,,	Mechanical	1	Max W.P. 20 Kg cm² Test press 26 ,, ,, Test date 1969
-90-01	Electric motor (fan) for cooler. A.S.E.A. No 110-016-C: 0·33 h.p. Type MT 71A 14-4	,,	Electric (minor)	1	415 volts Y: 1400 r.p.m. 3 ph. 50 ~: 0·25 kW 1 amp.
-66-01	Air receiver (air comp^r) Hudson & Brown Ltd: No HB 9775 To B.S.S. 487/60 GR D	,,	Pressure vessel	1	W.P. 100-125 P.S.I. HYD Test Press 195 P.S.I. Date of test 3-9-69

2.1 An inventory sheet

5. Instruments and instrumentation systems.
6. Pressure vessels, receivers, gas holders.
7. Lifting gear, lifting machines, lifting tackle, hoists, lifts, jacks, cranes, slings.
8. Machine tools – lathes, shapers, grinders.
9. Fire-fighting services – detection systems, alarm systems, fire pumps, extinguishers, appliances.
10. Factory and office services – heating, ventilation, hot and cold water.
11. Mobile plant – trucks, dumpers, small generator sets, mobile pumps, compressors.
12. Major spares – the inventory should also include major spares, i.e. rolls, gear boxes, heat exchanger tube nests, etc., which may be taken out of one machine, reconditioned, returned to stores and then used on another machine. These items will probably be in storage, but could also be in the workshops undergoing reconditioning.

Priority Rating: This rating ranging from 1 down to 5 indicates, in descending order, the relative importance of that particular item within the production process.

No. 1 Rating – Applied to all items whose efficient operation is vital to the production process. A failure of one of these items would immediately affect or halt production; alternatively, it would create a safety hazard.

No. 2 Rating – The failure of one of these items would not immediately affect production but could do so within a very short space of time.

No. 3 and 4 Rating – Similar to rating No. 2 but in descending order of importance.

No. 5 Rating – Plant whose failure would not affect production, or create a safety hazard in any way whatsoever. This usually applies to non-productive items of equipment.

The rating of each piece of equipment will depend upon the factors peculiar to the situation, but to assist the persons compiling the inventory, a standard for each rating should be defined and adopted.

Remarks: Any relevant notes that will assist in constructing the plan.

Note: Where office mechanization is in use identification symbol, location, type and priority rating is expressed in terms of a numerical code.

Having completed the inventory, it is now possible to decide on the items that are to be included in the maintenance plan. Priority will be given to items having a No. 1 Rating, as the efficient functioning of these has a direct and immediate influence upon production. But due allowance must be made for the inclusion of items which, although are not directly connected with the productive process, can exert a considerable influence upon it by means of outside pressures. For example, there may be no reason technically why production cannot proceed in the absence of factory space heating in winter or ventilation in summer, but for human comfort and satisfactory working conditions these services may be essential. Their failure would lower human activity and efficiency while in extreme cases a labour dispute, stoppage or walk-out could result.

2.3. Identification of plant and equipment

Identification symbols

When an inventory has been compiled it is essential that every item is positively identified so there can be no possible doubt as to which one is being referred to. This means that each item must be assigned an exclusive symbol, and by using the same symbol in documentation it is possible, with certainty, to relate instructions, records, job cards, specifications, reports, etc., to the item. The symbol need not be merely for the identification of each item; it is possible to design and utilize it to locate rapidly any item listed in the inventory, and also to provide management with a basis for cost allocation. These additional services may be rather elaborate for a small workshop containing a few different machines where there is no problem identifying and proceeding straight to the machine concerned. However, the situation is completely different in a factory containing numerous machines where many of them may be similar or even identical. In the absence of guidance, considerable time can be wasted in locating a particular machine.

Again, in a small workshop, the allocation of maintenance costs between the various units may not be necessary, but in a large organization the precise allocation and comparison of costs, either departmentally or between the various machines, is of the utmost importance.

Thus, summarizing the main reasons why each item should be positively identified:

1. So there can be no mistake about which particular item is being referred to.
2. To identify and relate the respective items with their relevant documents, and vice-versa.
3. To locate the whereabouts of the respective items within the site.
4. To indicate the department, section, group or type of item for cost allocation purposes.

These requirements can be met by codes made up from colours, shapes, patterns, names, letters, numbers or combinations of any of them, the choice depending upon the circumstances. But whatever symbols are used and whatever system is followed, there are certain conditions to be observed if it is to prove satisfactory:

1. The system must be designed logically.
2. It must be capable of accommodating changes in equipment without disorganizing the system.
3. Each item must have a unique symbol; this should be the only one by which it is known and identified.
4. The symbol must be one that is easily understood, recognized, and also be capable of reproduction on the particular item and all its relevant documents.
5. The same system of coding must be used throughout the factory.

Whilst codes can be constructed from a number of different symbol forms, for practical reasons the choice is usually restricted to names, letters or numbers. The disadvantages of colours, shapes and patterns for the identification of items will become apparent as the chapter proceeds, but these forms, when used in conjunction with other systems, can be extremely useful as visual aids to assist the easy recognition of the different plant or area groupings within the site.

Consider the alternatives – names, letters and numbers – for identification. All items of plant when first installed usually have an official name and title, but during the course of time, usage and the trend in modern terminology, names gradually are changed, often becoming unrecognizable with the original. Frequently, the same piece of equipment is known by different names and, except when dealing with only a very few items, the systematic filing of documents can be difficult and confusing, while the marking of equipment with its full descriptive name can be cumbersome.

The use of letters in combination with numbers is better. Every identification symbol comprises a letter and a number. Each department is allocated a distinctive prefix letter, usually the first letter of its departmental name. An exclusive number is then allocated to each item within that department. The identification symbol is then built up by the departmental prefix letter followed by the item number (e.g. M 42). By this means, plant can be recognized in its departmental group and the paperwork can be conveniently filed under its alphabetical prefix. Care must be taken to ensure that each department is allocated its own exclusive prefix letter, this applies particularly to departments having the same initial letter. The following is an example of this type of coding:

Prefix letter for foundry – F
Prefix letter for toolroom – T
Prefix letter for machine shop – M
Prefix letter for boiler house – B

Consider identification symbol M 42:

Departmental prefix letter – M
Item number within the department – 42
Identification symbol M 42 refers to Item No. 42 in the machine shop.

There are, however, certain disadvantages with this method; the symbol indicates only limited information, and letters of the alphabet are difficult to adapt to office mechanization systems. The implication of this latter fact should be appreciated in considering future developments. Increasing use is made of machines, and computer data processing systems for sorting and tabulating information. The information and data are fed into the machines in the form of a numerical code, so if maintenance data, including plant identification symbols, are to be accepted by the machines, they must also be in a completely numerical code. There are many systems that will fulfil this criterion, but each system must be adapted to suit individual requirements.

The following example illustrates a basic approach for constructing a numerical identification code for machine tools contained in a large factory divided into various departments. In this case, the object is to provide every machine with its own unique symbol and to indicate its type and location for data processing purposes. Each identification symbol will be made up of a six digit number. Reading the number from left to right:

The first two digits indicate the location of the machine, i.e. the department. The next two digits indicate the type of the machine, i.e. lathe, shaper, planer, etc. The last two digits indicate the machine number within its particular type group, i.e. No. 1 Lathe, No. 2 Lathe, No. 3 Lathe, etc.

Note: A two digit number 01 to 99 can accommodate a series of 99 different symbols but where a greater number are required a three digit series 001 to 999 must be used. *A location index* must now be drawn up. Every department is allocated an individual number within the series 01 to 99.

Location index:

01	Machine shop
02	Welding shop
03	Fitting shop
04	Pattern shop
05	Foundry
06	Press shop
07	Boiler house
08	Compressor room
09	Maintenance workshop

Similarly, a *machine type index* must also be drawn up. Each type of machine is allocated an individual number within the series 01 to 99.

Machine type index:

01	Lathe
02	Universal milling machine
03	Shaper

04 Planer
05 Surface grinder
06 Cylindrical grinder
07 Drill
etc.

Consider identification symbol 01 – 03 – 05:

Location	Machine type	No. of machine within the group
↓	↓	↓
01	03	05
↓	↓	↓
Machine shop	Shaper	No. 5

This identification symbol 01–03–05 immediately expresses the fact that it refers to No. 5 Shaper in the machine shop.

Although the example was concerned with the identification of complete machines, the same reasoning can be applied equally, in further detail, to identify major components of these machines – gear boxes, lubrication pumps, filters and hydraulic rams.

In this case, an index of machine components must be drawn up so it can be incorporated into the identification symbol.

Frequently, it is desirable for the identification symbol to indicate not only the type of machine but also its size. This can be accommodated by allocating a series of numbers to that particular type of machine. Each different number in the series then represents a certain size of machine. For example, assume that the machine type index has allocated the series 90 to 95 inclusive to electric motors, then

90 signifies an electric motor under 1 kW
91 signifies an electric motor 1 to 5 kW
92 signifies an electric motor 6 to 10 kW
93 signifies an electric motor 11 to 25 kW
94 signifies an electric motor 26 to 50 kW
95 signifies an electric motor over 50 kW

Consider the identification symbol 08–94–01

Location	Machine type and size	No. of machine within group
↓	↓	↓
08	94	01
↓	↓	↓
Compressor room –	electric motor 26 to 50 kW –	No. 1

The possibilities offered by this system are considerable; symbols can be built up to represent a very wide range of information peculiar to any specific component, machine, plant or process.

Marking the equipment

When an item of equipment has been allocated an identification symbol, it should be marked clearly and, where possible, by a method that is standard throughout the factory. By using a distinctive but standardized method of marking, we can distinguish easily the identification symbol from any other number of nameplate the item might bear.

There are many methods and variations that can be used, each suited to particular situations; due consideration must be given to the type, size and location of the equipment. A method that is ideal for small portable bench equipment may be totally inadequate for heavy plant operating out of doors in difficult conditions. No single method of marking will accommodate widely differing characteristics with complete satisfaction; therefore it is better to adapt the method to suit the equipment. This is quite in order provided that the methods, when selected, are then standardized for the respective categories of equipment, and are known to all concerned.

Although marking can be effected in various ways, certain features should be common:

(a) Easily visible – distinctive.
(b) Permanent – durable – will not deteriorate, rub off, wear off or fall off.
(c) Will not materially affect, impair or stress the item or part to which it is attached.
(d) Will not cause obstruction, inconvenience or create a hazard.
(e) Cheap and easily obtainable.
(f) Easily and quickly affixed.
(g) Neat in appearance.

So far, we have considered the identification and marking of equipment for planned maintenance purposes only, but there are many other reasons for positively identifying items of plant. Some are motivated purely by operational or administrative needs and may be applied solely at the discretion of management. Identification of the company's assets, references for stores and spares, references for valves, controls or switches to clarify instructional or operational manuals, are but a few examples.

There are other instances in which the positive identification and marking of equipment is a statutory regulation covered by the Factories Acts and other relevant legislation:

Power presses – Every power press and safety device must be marked clearly with some means of identification.

Steam boilers – Every boiler for which a test certificate has been issued must be so marked that it can be identified as the boiler referred to in the certificate. If there is more than one boiler in a factory, each must have a plate bearing a distinguishing number.

Air receivers – If more than one air receiver is in use in the factory, each must bear an easily visible identification mark.

Chains, ropes and lifting gear – Must be marked with some means of identification and also, with certain exceptions, be marked with the safe working load.

Cranes – Every crane must be plainly marked with the safe working load and a means of identification.

Blasting of castings (protective clothing) – Each protective helmet must carry a distinguishing mark indicating the person by whom it is intended to be used.

Means of identification

The Factories Acts require that a means of identification and the safe working load to be marked on all lifting tackle.

Chaintabs are rugged steel marking and identification labels which can be easily and permanently fitted to slings, chains, chain-blocks, rope-blocks, winches, cranes and all types of lifting tackle. They can also be used for identifying valves, portable equipment, keys, etc. (see Fig. 2.2a).

The tabs are easily visible, do not obstruct the use of the equipment, are not affected by heat treatment and do not constitute a stress raiser. They are available in five sizes.

After marking a Chaintab with the appropriate SWL and identification number, the ring is opened spirally wide enough to be passed over the selected part of the tackle, and the tongue is then hammered back into its location.

Ropetabs are tough flexible plastic labels of 'Y' section, used for marking ropes (see Fig. 2.2b). The tail of the Y, which is 51 mm (2 in) square projects from the rope to form the marking label. This label and the two 'arms' of the Y are each gripped securely between two adjacent strands of the rope. The arms continue at each end of the tab as a 'V' section, and make the device virtually impossible to displace from the rope in use, although it is quite easily inserted, or removed if required. The relevant identification number and SWL can be written in permanently by means of a hot wire point or impressed with type or letter punches heated to about 150 °C.

Cabletabs are for identifying wire ropes, electrical conductors, welding cables, etc. They are easily applied to endless cables. As Fig. 2.2c shows, a sherardized split steel ring fits over a plastic split sleeve of internal diameter equal to the size of the cable to be marked.

In the case of a cable obstructed at the ends, the plastic sleeve is slipped onto the cable. The steel ring is placed adjacently and moved along to cover the plastic sleeve and then squeezed to grip the cable tightly. The steel ring is stamped with the appropriate information, identification and SWL prior to fitting.

Both the plastic sleeve of the Cabletab and the plastic Ropetab are obtainable in several different colours which can be used as guides to indicate different inspection periods, different departments, etc. The sleeves are available in nine sizes – $\frac{1}{4}$ in (6 mm) to $\frac{3}{4}$ in (18 mm) by $\frac{1}{16}$ in (1·5 mm) increments.

Guardtabs. The Power Press Safety Regulations require that 'every

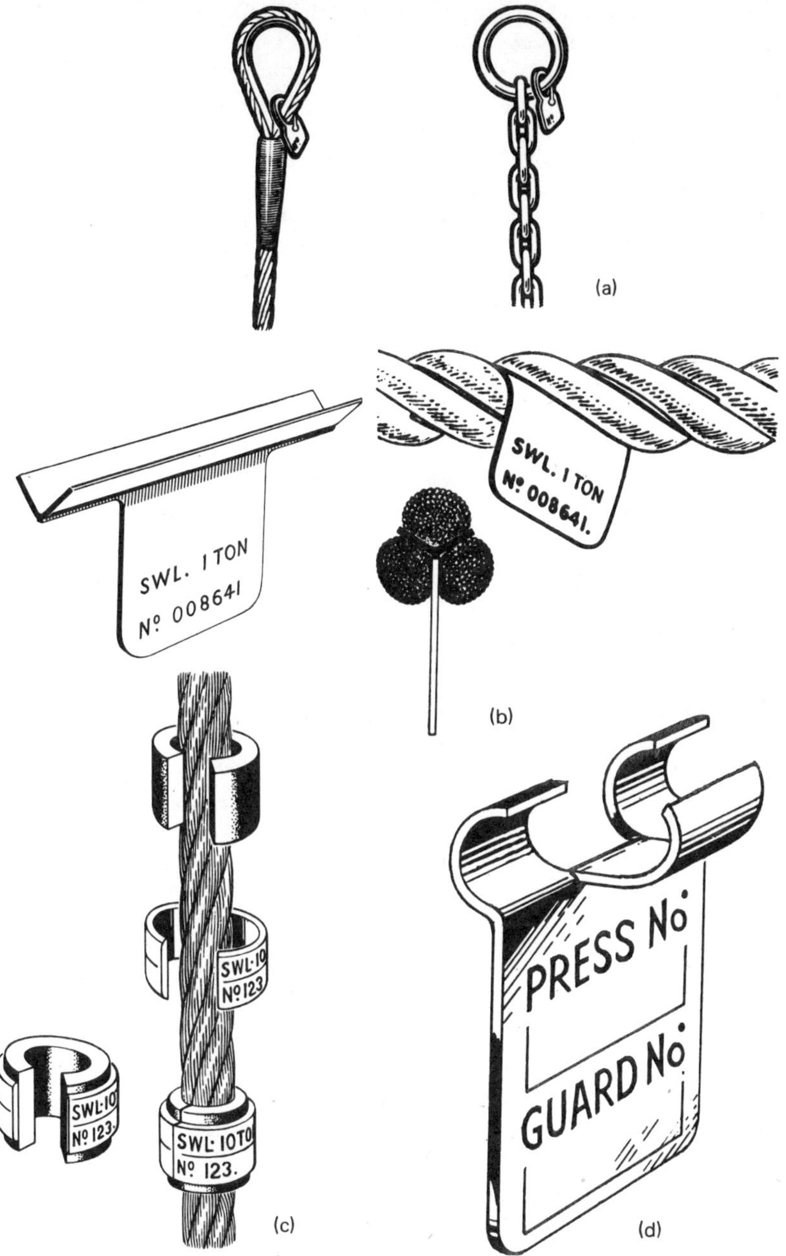

2.2 Identification labels (not to scale)
(a) Chaintabs (b) Ropetabs (c) Cabletabs (d) Guardtabs

power press and every safety device shall be distinctively and plainly marked'. Guardtabs are strong metal labels designed to be fixed easily to the mesh or framing of the panels of press guards by simply closing the claws on to a suitable member by squeezing or hammering (see Fig. 2.2d).

2.4. The facility register

The facility register is a record of facilities, including technical details about each.

Items that are to be included in the maintenance plan should be selected from the inventory and the 'vital statistics' about each entered in the Facility Register. The register records in a convenient and concise form, all the essential details about each item so that the information is readily available as a standard reference:

to confirm the original specification, performance,
to confirm manufacturers recommended limits, fits, tolerances,
to assist the ordering of correct spares and replacements, and
to provide the necessary information when planning the movement, relocation, access, safe floor loading, and layour of plant.

The value of this collection of data will be appreciated by those who have spent considerable time and trouble, often fruitlessly, searching for similar elusive details.

Identification No: Location:

Type of Facility:

Manufacturer: Date of Manufacture:

Serial No: Specification: Size: Model:

Capacity: Speed:

Total Weight: Power and Service Requirements:

Connection Details: Foundation Details: Overall Dimensions:

Headroom, clearance and access dimensions:
(a) For the withdrawal and maintenance of components.
(b) For manoeuvrability through restricted openings, doorways, passages, gangways, etc.

Reference drawing numbers:

Reference numbers of service manuals:

Interchangeability with other units:

Before the task of compiling the register can proceed it is necessary to decide:

what information will be recorded,
how the information will be recorded,
how the recorded information will be stored, and
how the register will be formed.

The nature of the data to be recorded will depend upon the type of facility. Nameplate details and information from manufacturers' literature are a starting point to which further information may be added when available. A general basis for most mechanical items is set out opposite.

In the case of electric motors the basic information usually recorded includes:

Identification No: Location:

Manufacturer: Serial No: Date of Manufacture:

Type: Specification: Rating: Frame Size:

Power/kW: Winding: Speed: Total Weight:

Voltage: Current: Phase: Frequency:

Shaft details: Diameter, length, keyway, height:

Bearings: D.E., N.D.E., Lubrication:

H.D. bolts: Diameter and centres

Reference Drawing Numbers:

Interchangeability with other motors

Identification number of associated starter gear:

When the facilities are both simple and straightforward these elementary details may suffice, but when complex and/or high precision equipment is concerned, which involves closely designed limits, fits and tolerances (dimensional, thermal, electric, electronic), it becomes necessary to supplement these basic data with specialized information. Figure 2.3 is an extract from a manufacturer's service manual for an oil engine and illustrates clearly the types of specialized data.

Information on items listed in the inventory as Major Spares should include not only the above relevant details but should also indicate their current location, condition and serviceability. Sufficient space should be

ENGINE CLEARANCE DATA

Details between which wear occurs	*Designed Max. and Min. Clearance*	*Maximum allowable Clearance*	*Remarks*
Crankshaft, end journal Main Bearing, end	.0025" to .0035" .064 m/m .089 m/m	.006" .152 m/m	Max. ovality of journals .003" (.076 m/m) Min. clearances may be reduced .0005" .012 m/m by 'Nip' when fitting bearings
Crankshaft, intermediate journal Main Bearing, Intermediate	.002" to .005" .051 m/m .127 m/m	.007" .178 m/m	
Crankshaft End Float	1 Cyl. Engine .002" to .007" .051 m/m .178 m/m Multi Cyl. Engine .003" to .005" .076 m/m .127 m/m	.015" .38 m/m .010 .254 m/m	
Crankpin Large End Bearing	.0015" to .004" .038 m/m .102 m/m	.006" .152 m/m	Max. ovality of crankpin .003" .076 m/m Min. clearance may be reduced .0005" .012 m/m by 'Nip' when fitting bearings
Liner Piston Body (Cast Iron)	.004" to .0055" .102 m/m to .14 m/m	.008 (.203 m/m) on unworn section of liner	Maximum Liner wear .015" (.38 m/m) The Liner wear is more important because piston body wear is usually negligible
Liner Piston Body (Alloy)	.0045" to .013" .114 m/m .33 m/m Piston oval and tapered	.009 to .016 .23 m/m to .41 m/m on unworn section of liner	
Gudgeon Pin Boss Gudgeon Pin	.0005" to .0015" (C.I.) .012 m/m .038 m/m .0005 .012 m/m interference to .0005" .012 m/m clearance (alloy)	.003" C.I. .076 m/m .0015" alloy .038 m/m	
Gudgeon Pin Small End Bush	.0015" to .0025" .038 m/m .064 m/m	.004" .102 m/m	
Piston Ring Groove (Pressure) Piston Ring	.001" to .003" .0254 m/m .076 m/m	.008" .203 m/m	Grooves can be opened out for overwidth rings
Piston Ring, gap in position	.010" to .015" .254 m/m .38 m/m	.030" .76 m/m	Always check gap on unworn portion of Liner
Piston, scraper rings	SIMILAR TO PRESSURE RINGS		

2.3 Extract from a service manual

allocated on the document to record the recurring cyclic changes in plant movement, wear and tear, restoration and storage that can occur.

Manufacturers' service manuals and reference drawings can be a valuable source of information. Each should be given a reference number and filed away safely, only the reference numbers need then be noted in the register. When literature of this type forms an integral part of the register, it must be available when required.

All this information must now be set out and presented in a manner best suited to the particular circumstances.

When only a few pieces of equipment are involved, simple office duplicated forms contained in loose-leaf binders or cards filed in drawers (Fig. 2.4) may be adequate. For more extensive applications, it is probably better to obtain a complete tailored system from one of the specialist suppliers. The majority of these proprietory systems rely upon the data being written or typed on printed loose sheets or cards – visible records (Fig. 2.5). They are available in various sizes and colours, either designed to the customer's own specific requirements or chosen from the suppliers standard stock range.

Registers are compiled by filing the sheets into books (Fig. 2.6) and the cards into flat trays (Fig. 2.7), a page or tray holding several dozen overlapping documents. Each document is so located that only 9·5 mm ($\frac{3}{8}$ in) of its lower edge is visible, this visible portion bears the identification symbol of the item concerned. By glancing down the appropriate page, or tray, the required document can be seen and located easily (Fig. 2.8).

Normally, only one sheet or card is allocated to each item of equipment, but where one card is insufficient to contain all the necessary data supplementary sheets can be included. If a particular item of plant contains major components or sub-assemblies – rolls, heat exchangers, gear boxes, pumps – that are transferable or capable of individual replacement, it is usually more convenient to allocate a separate, supplementary sheet to each of these sub-units, these additional sheets being attached to the main record sheet within the register. Each sheet or card must bear the identification symbol of the item or part to which it refers.

The retrieval of information from conventional visible records systems is a manual task. Even with a moderately-sized inventory the searching and extraction of specific items of data, for example, all electric motors of a particular power and speed, all equipment installed in a particular department, all machines of a particular type, size, and power requirement listed in the register or in a particular department, all pumps of a specific capacity or all machines over a certain age, can prove time consuming, tedious and subject to mistakes and omissions. A fully computerized data storage and retrieval system would greatly facilitate the task, but such a sophisticated system is extremely costly and out of reach to all but the largest of organizations. Even when computers are installed, the time needed and cost involved for programming and processing, plus

IDENTIFICATION Nº		LOCATION
MANUFACTURER		SERIAL Nº YEAR
K.W.	R.P.M.	TYPE
VOLTS CURRENT	PHASE WINDING	FREQUENCY RATING
FRAME SIZE	MOUNTING	WEIGHT
SHAFT DIA KEYWAY	LENGTH	CENTRE HEIGHT
BEARINGS–D.E.	N.D.E.	LUBRICANT
ASSOCIATED STARTER GEAR		
ASSOCIATED EQUIPMENT		
INTERCHANGEABLE WITH MOTOR Nº		

FACILITY CARDS FILED IN DRAWERS.
THE RESPECTIVE CATEGORIES
SEPARATED BY DIVIDERS

01–04

05–09

2.4 A facility register and storage thereof

MECHANICAL GROUP 3 - MISCELLANEOUS

DESCRIPTION

SPEC. NO.

ORDER NO.

TYPE

MANUFACTURER

ADDRESS

TEL. NO.

COMPONENTS	NO. FITTED	SERVICING REQUIRED	LUB. REQUIRED	REMARKS

SPECIAL INFORMATION

Kalamazoo
375479-611

TITLE BELOW THIS LINE

DESCRIPTION / SERVICE	LOCATION	SERIAL NO.	CODE

(front)

DRAWINGS

ITEM	DRAWING NO.	INDEX REF.	ITEM	DRAWING NO.	INDEX REF.

SPARES

ITEM	MAKERS NO.	SUPPLIER	NO. HELD	BIN. REF.	ITEM	MAKERS NO.	SUPPLIER	NO. HELD	BIN. REF.

SPECIAL TOOLS

ITEM	MAKERS NO.	PURPOSE	WHERE HELD	ITEM	MAKERS NO.	PURPOSE	WHERE HELD

Kalamazoo mechanical card (back)

2.5 Examples of facility record sheets for visible records systems

PLANT AND MACHINERY RECORD (MECHANICAL)

MAKER
MAKERS NO.
COST
SUPPLIER
NEW/SECOND HAND
DATE
DESCRIPTION
SPARES
DRIVE
INSURANCE
GUARANTEE
POLICY NO.
RENEWABLE

Date	Location	Ref.	Notes

EXPECTED LIFE ___ YEARS
Weight
Area
SERVICES REQUIRED
A.C. Single
A.C. Three
D.C.
Gas
Water
Vacuum
Air at ___ lbs. Square Inch
Steam at ___ lbs. Square Inch
Hydraulic Oil
Dust Extraction
Type of Foundation
Lifting Tackle
Coolant
Lubricants

Drive	H.P.	Speed	Ref.

Drawing Nos.

Kalamazoo 1511-611
TITLE BELOW THIS LINE

OOA	LOCATION	DESCRIPTION	CATEGORY	NUMBER

INSPECTIONS J F M A M J JY. A S O N D

Kalamazoo mechanical card

PLANT RECORD (ELECTRICAL) 1

H.P.
R.P.M.
STATOR VOLTS
AMPS. FL.
MAKE
ST. TORQUE % MAX.
ROTOR VOLTS
AMPS. FL.
SERIAL NO.
VENTILATION
D.E. BEARING
TYPE
SHAFT FITTED WITH
NDE ..
FRAME SIZE
SIZE
SHAFT DIAM.
WIRING DIAG. NO.
OUTLINE DWG. NO.
WEIGHT
DATE COMMISSIONED
TESTED BY
INSULATION RESISTANCE (MEGOHMS). BETWEEN PHASES
TO EARTH

SPARES AVAILABLE	LUBRICATION	REMARKS
	GREASE/OIL	
	MAKE	
	GRADE	
	TOP UP	
	CHANGE	

RED
YELLOW
GREEN
BLUE

Kalamazoo 463634-611
TITLE BELOW THIS LINE

MOTOR DRIVING	LOCATION	WORKS NO.

1 2 3 4 5 6 7 8 9 10 11 12

Kalamazoo electrical card

2.5 (continued)

↓ Commence typing here

In order for the titles to be VISIBLE, they must be typed within ½" below perforation—as close to the perforation as possible. After typing front and back, tear off perforated strip.

INSTRUMENT DATA

TYPE
CONSTANTS
RANGE INPUT OUTPUT
MOUNTING FINISH
ACTION PRESSURE
AUX. SERVICES
MATERIAL
CHART DRIVE SPEED
AUX. EQUIPMENT

FLOW ELEMENT DATA

TYPE
NOM-SIZE FLANGED
THROAT SIZE MATL.
OUTPUT RANGE
PRESSURE
DIFF. PRESSURE
FLUID
AUX. SERVICES
AUX. EQUIPMENT

VALVE DATA

CONNECTION
SIZE FACE
MATERIAL
FLUID
PRESSURE TRIM
TRIM SIZE ACTION
MOTOR TYPE No.
DIAPH. DIA. AREA
SENSITIVITY
REVERSIBLE
PACKING

COUPLING DATA

MAKE
SIZE
DRIVING HALF BORE KEYWAY
DRIVEN HALF BORE KEYWAY

BEARING DATA

DRIVE END TYPE
MAKE EQUIV.
No. No.
DRIVEN END TYPE
MAKE EQUIV.
No. No.

DRIVE DATA

TYPE
PATIO H.P. CRS.
DRIVING PULLEY R.P.M. P.C.D.
BORE FACE GROOVES
DRIVEN PULLEY R.P.M. P.C.D.
BORE FACE GROOVES
ROPES, SIZE
No. LENGTH
DRIVING SPROCKET R.P.M. P.C.D.
BORE No. OF TEETH
DRIVEN SPROCKET R.P.M. P.C.D.
BORE No. OF TEETH
CHAIN SIZE No. OF LINKS

RECOMMENDED LUBRICANTS

DRAWING Nos.

(front)

RONEODEX VISIBLE RECORDS, RONEO LTD., 17 SOUTHAMPTON ROW, W.C.1

RONEO LTD. RDX. No. 078425A/G.D

DESCRIPTION

MAKER SUPPLIER

ORDER No. DATE
SERIAL No. REF. No.
TYPE
TYPE OF DRIVE

YEAR OF MANUFACTURE
MODEL No.
CLASS
H.P.

INSTALLATION DATE
INVOICE PRICE
B.S.S.
R.P.M.

PUMP AND COMPRESSOR DATA

HEAD SUCTION DELIVERY
DIA. SUCTION DELIVERY
G.P.M. IMP. U.S.A.
IMPELLER DIA % CONSIST
HAND
No. OF CYLINDERS
BORE TROKE
WORKING PRESSURE LBS./SQ. IN.
MAX. PRESSURE LBS./SQ. IN.
OIL PRESSURE LBS./SQ. IN.
AIR DISPLACEMENT CU. FT./MIN.
WATER GAUGE INS.

GEAR BOX DATA

CENTRES RATIO
RPM. INPUT OUTPUT
HAND

CHEST AND TANK DATA

CAPACITY CU. FT. GALLS.
DIMENSIONS
CONSTRUCTION
LINING
WORKING PRESSURE LBS./SQ. IN.
TEST PRESSURE LBS./SQ. IN.
AGITATION TYPE

MOTOR AND GENERATOR DATA

CURRENT FRAME SIZE
VOLTS AMPS
CYCLES PHASES
CONNECTED WOUND
RATING
ROTOR VOLTS ROTOR AMPS

STARTER DATA

SUPPLY
CONTROL VOLTAGE AMMETER
FULL LOAD CURRENT C. T. RATIO
PROTECTION RATING
MOUNTING

SPECIAL DETAILS

DESCRIPTION PLANT NO

↑ Commence typing here

In order for the titles to be VISIBLE, they must be typed within ½" above perforation—as close to the perforation as possible. After typing front and back, tear off perforated strip.

Roneodex card (back)
2.5 (continued)

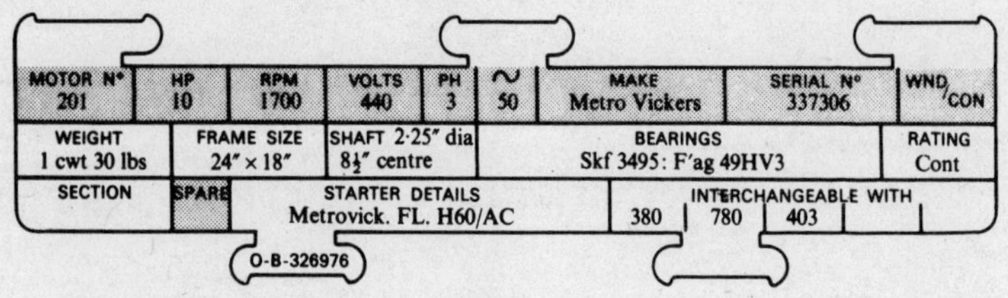

Kalamazoo Strip Index when applied as a facility record (see section 5.2)

2.5 (continued)

administrative difficulties are enough to discourage their use for all except major projects.

Many of the advantages without the disadvantages of a computerized system are offered by punched edge cards. Each item of equipment is allocated a separate card, the data to be stored are punched in code around the edges of it (Fig. 2.9).

There are two methods of indicating which of the programme slots should be punched out from the edge of the card:

1. Each digit, month, week number or other characteristic to be noted is allocated its own individual compartment along the edge of the card. This needs only one slot to be punched out to record each particular item of information (Fig. 2.10).
2. This is a more compact space-saving method which enables more data to be stored along the card edge. The digits 0 to 9 inclusive are formed into a pyramid.

   Each digit is located in the pyramid by the intersection of two coordinates. Thus the slots of the two co-ordinates must be punched out to indicate a particular digit (Fig. 2.11).

To indicate digit N° 0 cut out slots D and E
1 cut out slots C and E
2 cut out slots C and D
3 cut out slots B and E
4 cut out slots B and D
5 cut out slots B and C
6 cut out slots A and E
7 cut out slots A and D
8 cut out slots A and C
9 cut out slots A and B

The size of card is dictated by the number of facts to be recorded on any one card – the larger the card the more information that can be punched around its edges. The printed format of the cards is specifically designed to the user's requirements. Retrieval of the information is carried out with the aid of a small office machine (Fig. 2.12). Programme bars are positioned corresponding to the information to be extracted. The cards

(a) Desk strip index records

(b) Loose leaf records

2.6 The Kalamazoo Visible Records Systems

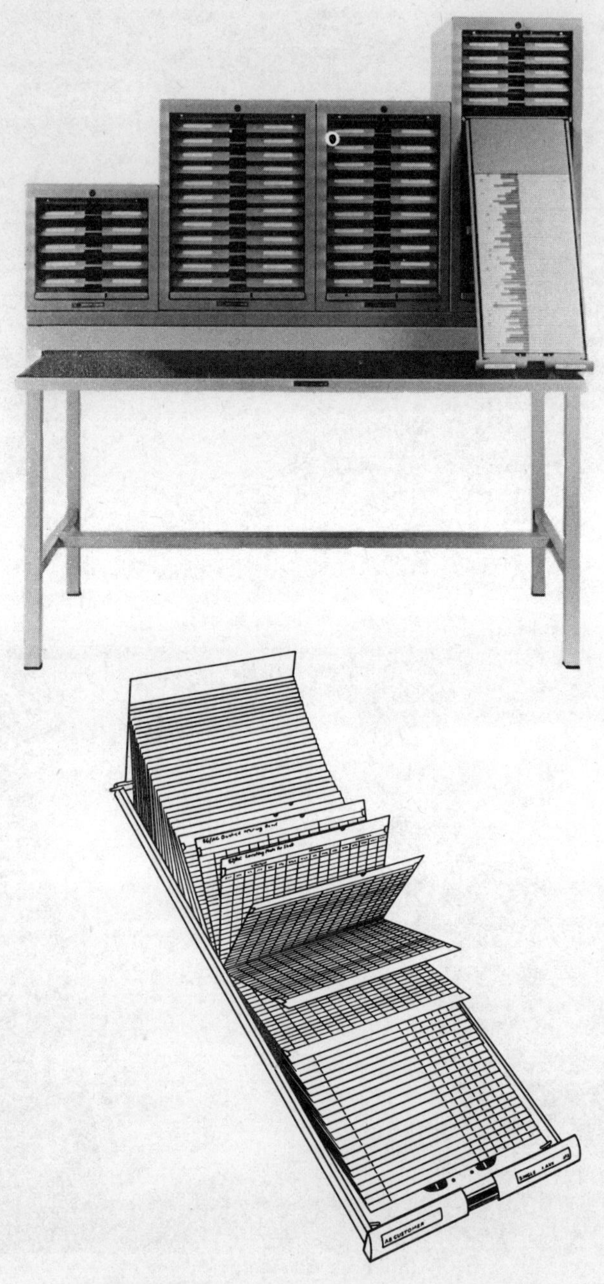

2.7 The Roneodex Visible Card Recording System

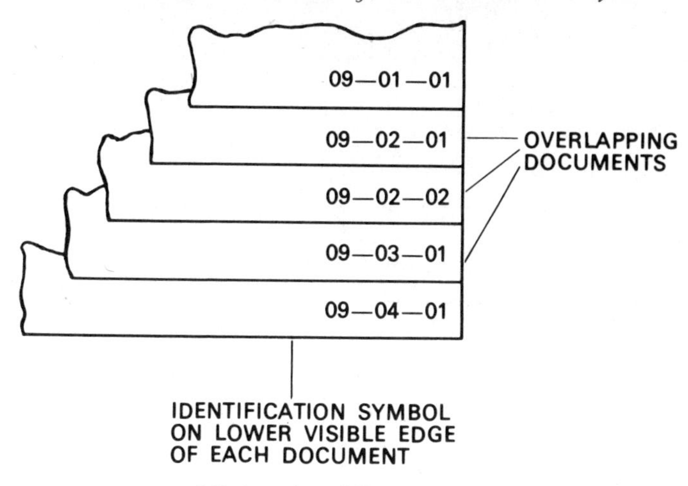

2.8 A section of filed sheets

are placed in the magazine of the machine which is vibrated electrically and the relevant cards drop to an offset position so they can be easily selected. It is claimed that by this method 5000 cards can be processed in as little as five minutes.

The choice of system should be influenced not only by the equipment involved but also the company's policy, documentation systems at present in use, and possible future standardization and rationalization of systems. Whatever system or method is finally adopted, it must be capable of accommodating changes and provide easy access to the recorded material.

For ease of reference, any collection of documents must be arranged in an orderly, systematic manner. This can be facilitated by maintaining separate sections of the register, or when the amount of equipment is sufficient, separate registers for each category of plant. The categories corresponding to the plant types are detailed in the inventory (section 2.2). Instant recognition of the document categories can be simplified by the use of different coloured stationery or by coloured flashes marked on the visible edge.

All registers and data must be kept up-to-date. This is especially pertinent to those sections or registers dealing with major spares; the records may well be consulted in cases of sudden breakdown to ascertain the condition and availability of particular spares, and determine the advisability of either repair or replacement.

Further details and descriptions of proprietory systems available are given in Chapter 5.

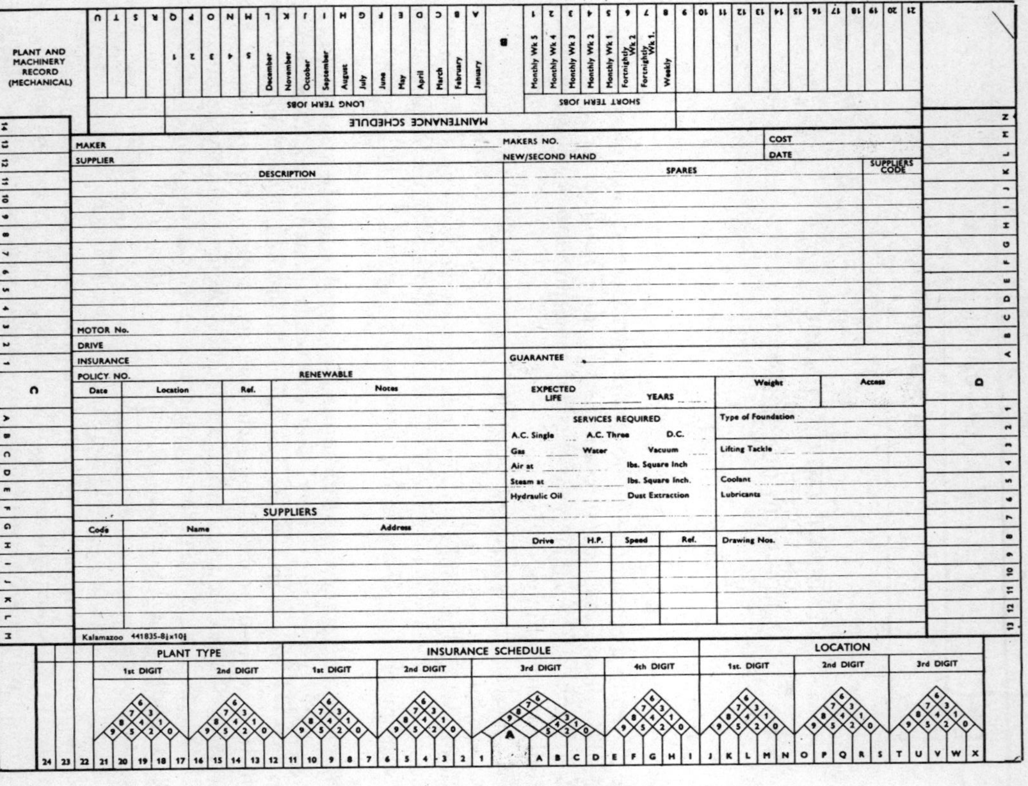

PLANT AND MACHINERY RECORD (MECHANICAL)

MAINTENANCE SCHEDULE

LONG TERM JOBS: January, February, March, April, May, June, July, August, September, October, November, December, 5, 4, 3, 2, 1

SHORT TERM JOBS: Monthly Wk 5, Monthly Wk 4, Monthly Wk 3, Monthly Wk 2, Monthly Wk 1, Fortnightly Wk 2, Fortnightly Wk 1, Weekly

MAKER
SUPPLIER
MAKERS NO.
NEW/SECOND HAND
COST
DATE

DESCRIPTION
SPARES
SUPPLIERS CODE

MOTOR No.
DRIVE
INSURANCE
POLICY NO.
GUARANTEE

RENEWABLE

Date	Location	Ref.	Notes

EXPECTED LIFE ____ YEARS
Weight
Access

SERVICES REQUIRED
A.C. Single ____ A.C. Three ____ D.C. ____
Gas ____ Water ____ Vacuum ____
Air at ____ lbs. Square Inch
Steam at ____ lbs. Square Inch.
Hydraulic Oil ____ Dust Extraction ____

Type of Foundation
Lifting Tackle
Coolant
Lubricants

SUPPLIERS

Code	Name	Address

Drive	H.P.	Speed	Ref.

Drawing Nos.

Kalamazoo 441835-8½x10½

PLANT TYPE: 1st DIGIT, 2nd DIGIT
INSURANCE SCHEDULE: 1st DIGIT, 2nd DIGIT, 3rd DIGIT, 4th DIGIT
LOCATION: 1st. DIGIT, 2nd DIGIT, 3rd DIGIT

2.9 A record card for a punched edge system

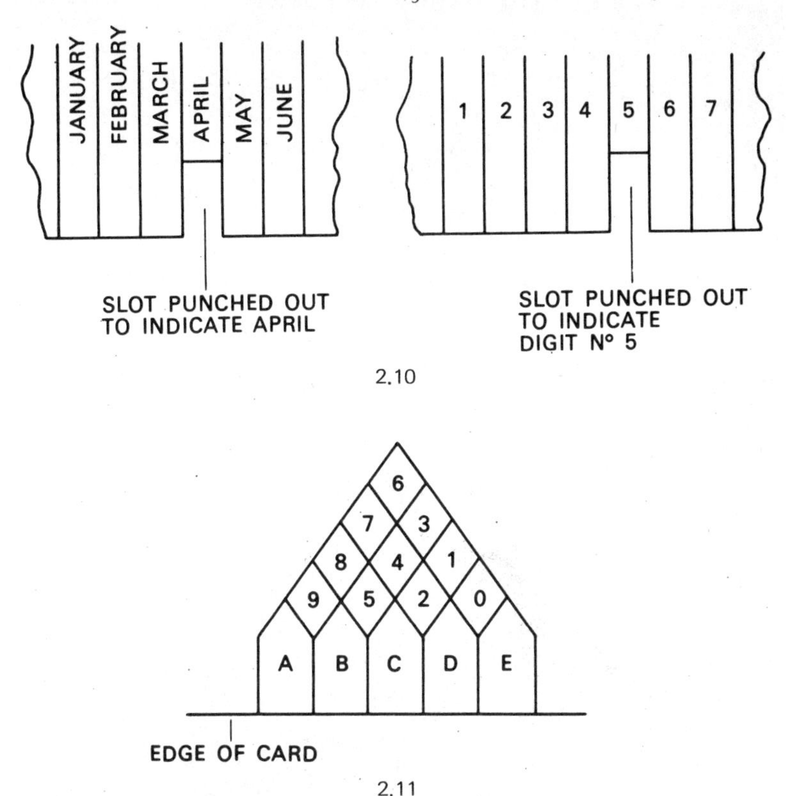

2.10

2.11

2.5. Maintenance schedule

The maintenance schedule is a comprehensive list of maintenance and its incidence.

Having gathered and tabulated within the facility register the information on the physical characteristics of the equipment to be maintained, we must establish *how it is to be maintained.*

For each facility included in the plan, a separate maintenance schedule must be compiled which sets out all the tasks – inspection, lubrication, adjustment, component replacement, overhaul – together with the frequencies that are considered necessary to maintain the facility efficiently. The maintenance policy for each facility should take into account its function and the influence it exerts over production. As an example, consider the equipment listed in the inventory (Fig. 2.1). The machine tools installed in the maintenance workshop have no direct or immediate effect on the production process, so they could well be placed on a breakdown maintenance basis with only lubrication and adjustment regularly carried

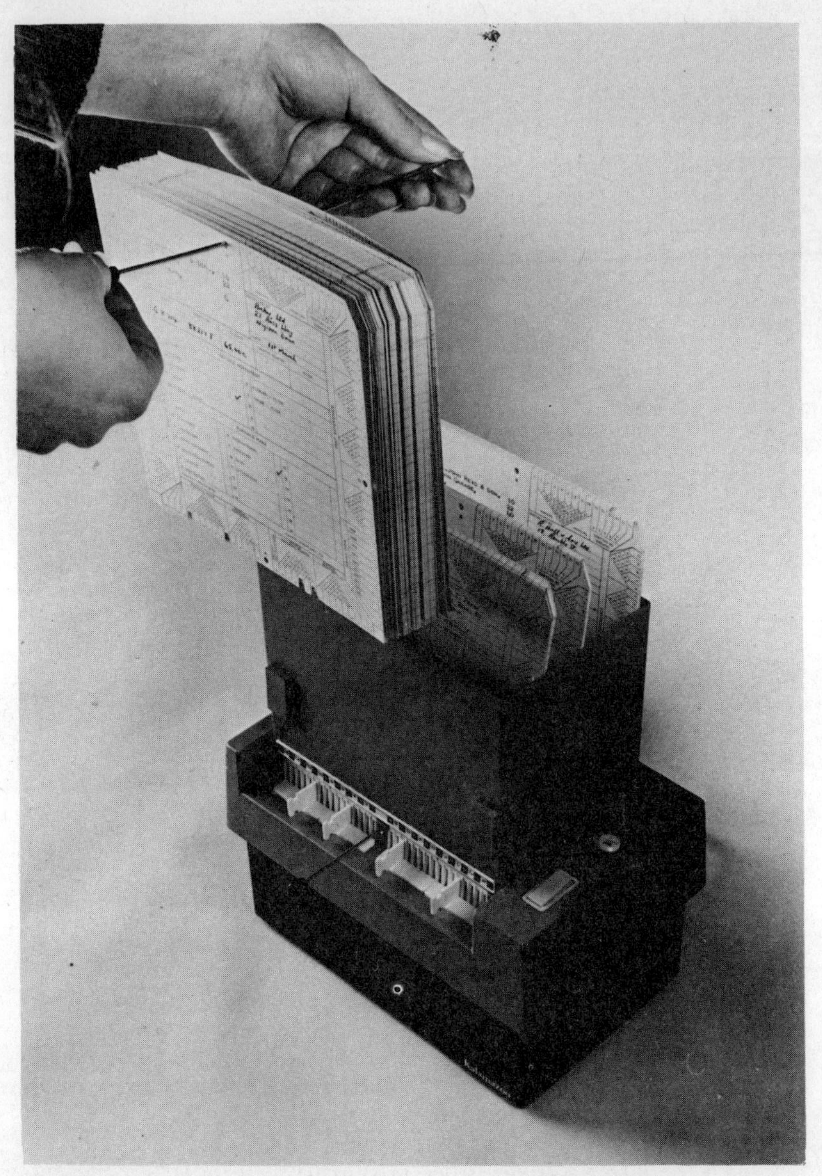

2.12 The Kalamazoo Factfinder — punched edge card system

Accuracy of punching on the cards is essential and a hand punch of 'precision tool' exactness is used. Cards are supplied in four different colours to give immediate recognition and are retained in a storage cabinet to protect their edges.

When a question is asked, programmed bars are inserted into the appropriate positions in the Factfinder to coincide with the punching on the cards which represents the required answer. The cards are placed in the magazine of the Factfinder and it is vibrated electrically, causing the cards with the selected data to drop and thus become offset slightly. Rods are inserted at the top and the bottom of the cards — the top rod brings away the unsorted cards, leaving the required cards in the magazine.

out. At the other end of the scale, the production process depends upon the air compressor – its failure in service would halt production and prove very costly in lost output. In this case, preventive maintenance is essential. Owing to the importance of the air compressor, we might aim for 100 per cent availability, but in practice high plant availability involves high preventive maintenance costs, costs which could out-weigh the financial gain of the extra production. The same philosophy could apply equally to the various components of a facility – components that are vital to its efficient operation would receive a higher level or different form of maintenance than the less important. The effect on production is not the only maintenance criterion, the cost of repair should the facility break down must also be considered – some repair jobs can be extremely expensive. A balance must be struck between all these factors so that the most economical level of maintenance is achieved.

Regular attention is essential for the efficient operation of most facilities. In the absence of any previous operational experience or knowledge of the particular equipment, the schedules must be based solely on personal judgement or, if available, on manufacturer's recommendations. Any information contained in manufacturer's literature should be taken only as a general guide and should be adapted to suit the relevant conditions. A maintenance schedule thus formulated, even if unsubstantiated, at least provides a starting point until factual information is available. When past records – plant log sheets, stores records, cost sheets, time cards – and operational experience are available, the schedule can be compiled on a more realistic basis. An analysis of the information extracted can help to establish:

- The facilities that most frequently fail.
- The components or units of each facility that most frequently fail.
- The type and frequencies of the failures.
- The result or effect of the various types of failure.
- The conditions under which failures most frequently occur.
- The length of time the facility, and/or its components can operate efficiently without attention.
- The effect of time and conditions upon efficiency and performance.
- The time and cost needed to carry out the various maintenance and repair jobs.
- The components and materials used for maintenance, repair or replacement.
- The tradesmen or contractors needed to carry out the work.
- The total cost of maintaining each facility.
- The allowable limits of wear, temperature rise, pressure drop, etc., if not detailed in the facility register.
- Constraints imposed upon the maintenance function by legal, operational or safety obligations.
- Safety practices and regulations to be observed.

Observance of the Factories Acts regarding the type and frequency of maintenance, examination and test.

Conditions when each facility is available for maintenance
- only when the whole process is shut down
- any time
- only when a defect or breakdown occurs.

Associated equipment or components that must be maintained simultaneously.

As the information accumulates and is studied, so the pattern of past events emerge, the past being a guide to the future. Collecting and analysing these data provide the opportunity to study closely the maintenance problems and needs of each facility. Before devising an overlong schedule of work which simply perpetuates previous tasks (plus a few more) to be carried out with even greater frequency, it would be more profitable to find ways and means of reducing, or better still eliminating, the need for maintenance. This approach is dealt with in more detail in other sections of the book, namely:

Designing out maintenance, section 3.1.
Method study, section 4.1
Maintenance aids, section 4.2

Having established the maintenance needs together with the methods, materials, replacements, tools, tradesmen and time required to accomplish them, we must determine the frequencies at which each of the various tasks or operations need to be carried out. These must be so balanced that they are neither too low (so the equipment is over maintained causing unnecessary maintenance work) nor too high (so the equipment is under-maintained resulting in premature failure).

Frequencies can be specified in terms of a calendar time scale, i.e. monthly, quarterly, annually, or in terms of operating time, cycle or mileage, i.e. running hours, operating cycles, miles travelled. A calendar scale has the advantage that it is possible to plan on a long term basis and spread the maintenance work load evenly over the whole year. It does not, however, allow for irregular or seasonal usage which could result in items being either over- or under-maintained. When maintenance is based on operational time or mileage, the total annual maintenance load is approximately proportional to plant usage, but as only short term planning is possible, programming difficulties and considerable fluctuations in work loading can arise. Failure can result from deterioration, corrosion, chemical attack and marine or algae growth, all of which can occur irrespective of plant usage. Wherever possible, maintenance should not be based solely on operational usage but should also incorporate an element of regular calendar time attention. Alternatively, the intervals between maintenance tasks can be specified either as a maximum time *or* a maximum usage, whichever occurs first.

Some facilities signal their own need for maintenance – a rising

SIX-MONTHLY INSPECTION

(1) Clean down external parts of motor and blow out air ducts. Check holding-down bolts for tightness.
(2) Clean out terminal box and check terminal connections for tightness and renew silica gel dryer if fitted.
(3) Check insulation resistance and continuity of windings with a 500 V megger and record reading before replacing terminal box cover.
(4) Check that the earthing strip is secure and there is earth continuity.
(5) Grease motor bearings with approved grease, using a gun.
(6) If the motor is fitted with sleeve bearings, drain oil from the bearings. Inspect bearing leads and record readings before reassembly.
(7) Clean out bearings with flushing oil and refill to the correct level, using the specified grade of oil.
(8) On a motor with sleeve bearings, check air gap readings at all points provided and record the results. If the motor coupling alignment is suspect this should be checked.

TWO-YEARLY INSPECTION

(1) Clean down motor exterior and blow out air ducts.
(2) Disconnect main motor main cables, alarms and auxiliary wiring and mark the cables for easy reconnection. Ensure that the cables are protected from moisture and mechanical damage.
(3) Disconnect motor from drive unit and transport to the approved working area or workshop, making sure that all packing pieces are identified and kept in a safe place.
(4) Draw coupling or pulley from the shaft and check that the keyway key and shaft are free from burrs. Check coupling for wear.
(5) Check sleeve bearings for wear, scores and measure oil clearance. Check that the lubrication holes and oilways are not blocked.
(6) Remove motor end covers.
(7) Check ball and roller bearings, if fitted, and renew bearings if required. If new bearings are not required degrease, clean and pre-oil existing bearings and repack with the approved grease when reassembling the motor.
(8) Remove rotor and check that the rotor bars and end shorting rings are free from cracks and are tight.
(9) Check that rotor laminations are tight and look for signs of rubbing between stator and rotor.
(10) Blow out stator winding with dry compressed air and clean stator winding of grease and dirt, using approved cleaning fluid.
(11) Examine stator winding for signs of loose or charred insulation and damaged binding tape. If the insulation appears to be in need of revarnishing this should be done with the approved varnish and the winding dried out in an oven or with fan-heaters, during which time the insulation resistance should be recorded.
(12) Check that stator laminations are free from burns and that the stator frame is clean and the air-ways are free.
(13) Reassemble the motor and refit the coupling.
(14) Replace the motor on the bed plate and realign to the coupling of the driven unit and record the results.
(15) Check the air gaps at all positions and record on a sketch.
(16) Reconnect all cables and test motor and cables for insulation resistance and continuity with a megger.
(17) Check that the terminal box is clean, ensure all gaskets are sound and renew the silica gel drier if necessary before replacing the terminal box covering.
(18) Check that the bearings are filled with the specified oil (on sleeve bearings only) and check that the motor is free to rotate by hand.
(19) Taking suitable precautions, run the motor uncoupled to check rotation and listen to the bearings with a "sonascope". If this is satisfactory, recouple the motor to the driven unit.

2.13 Maintenance schedule for a three-phase induction motor

pressure differential across a filter, a temperature increase, a pressure drop, a reduction in flow – each could be an advance warning. Provided these signs are not overlooked or ignored and no damage or failure will immediately occur, maintenance could be programmed for the first convenient occasion, but again planning will be on a short term basis.

Many facilities, e.g. electric motors, starters, motor vehicles, although supplied by different manufacturers are, similar in construction and maintenance needs. In such cases, it is not always necessary to prepare separate maintenance schedules for each unit – a standard schedule, universally applicable to all units within a particular classification, will usually suffice. Where local conditions must be taken into account it is often necessary to modify only the frequency of application. Figure 2.13 shows a typical schedule of work for a 30 kW, 415 V, 3-phase induction motor[16]. In favourable circumstances it might be necessary to inspect a similar motor only annually and completely overhaul it every three years, the work content remaining the same as that shown in the schedule.

To summarize

The schedule consists of an individual sheet, card or a set of sheets/cards for each facility. The sheets or cards indicate:

1. The name and identification number of the item of equipment.
2. The location of the item.
3. The reference number of the schedule (each schedule should be allocated an individual reference number).
4. Safety procedures to be followed (permit-to-work certificates to be obtained).
5. Detailed list of tasks to be carried out* (each task listed should be allocated an individual reference number).
6. The frequency at which each of the listed tasks must be carried out.
7. The tradesmen or other personnel required to undertake the respective tasks.

*Note: Each of the scheduled tasks must fulfil a need and make a positive contribution to the ultimate objective:

Will the facility be any more reliable or function any better as a result of the work? Will the inspection indicate the true condition or provide positive information?

Tasks that do not meet these conditions serve no useful purpose, they are obviously unnecessary and can be omitted from the schedule with no adverse effect.

The means must suit the end. The methods employed to accomplish the tasks must be effective, e.g. if the object of a particular task is to assess, with accuracy and certainty, the condition of a major component for further service, visual inspection and measurement will indicate only wear, scoring, ovality, etc. It will not reveal hidden cracks or flaws which also govern its acceptability. Outwardly, the component may appear flawless, only to break down later in service owing to an obscure fatigue crack or flaw. In this instance, visual inspection is not effective, only special equipment could have detected the fault.)

8. The time required to carry out the tasks.
9. Special tools or equipment required.
10. Materials, major components or replacements required.
11. When the facility is available for maintenance (running maintenance, shut down maintenance, etc.).
12. Associated equipment or facilities that should be maintained simultaneously.
13. Details of any contract maintenance.

From the completed schedules information should be extracted and acted upon so that everything is prepared. Spares, replacements, tools, special equipment and other necessary items should be ordered. Maintenance contractors should be contacted, pre-arrangements made and confirmed so that all parties know what is expected of them and are in agreement.

There are various ways of setting out a maintenance schedule, the most usual being similar to that illustrated in Fig. 2.14. This shows the maintenance required for a small air cooled diesel engine driving an electric generator. The various periodic tasks have been grouped according to their respective frequencies, each group being allocated its own service code:

'A service' to be carried out daily
'B service' to be carried out weekly
'C service' to be carried out every 2 months, etc., etc.

For reasons previously discussed (page 23), alphabetical service coding may not be acceptable, in which case numerical coding should be used – No. 1 service, No. 2 service, etc. A service code is a convenient means of summarizing and representing grouped tasks on documents and programmes, it is economical in space and recognized easily. The individual tasks are described only in their basic outline form – to have detailed each task in full would have made the schedule unreasonably long. However, the work content of each task can be ascertained easily as the schedule refers to the appropriate job specification or page in the service manual.

2.6. Job specification

The job specification is a document describing the work to be done.

Once the tasks necessary for the effective maintenance of each facility have been scheduled, the means of communicating the details to the person or persons who will effect the work must be considered.

If maintenance is to be effective then the planner must communicate the details clearly and precisely so that there can be no possible doubt about what is required. Seldom is it convenient or practical for him to explain the requirements every time the task is to be done; a means of instruction is necessary that will provide the link between the planner and the 'doer'.

MAINTENANCE SCHEDULE

Description of Facility "MBM" 15 kW Air Cooled Diesel Engine	Location: Pump House	Identification No of Facility 16–52–3
Ref Drawing Nos		Schedule Ref: No 52693
Service Manuals – Ref Nos 8234		Date Originated 25 April 1973
Associated Equipment Generator No 16–60–3		Modifications / Date

Item No.	Job Description	Availability for Maintenance*	Job Spec^n Applicable for each Item	Trade	Time Required for each Item	Remarks
	Daily – "A" Service					
1	Check level of fuel in service tank.	R		Lub^n Op^r	5 mins (items 1–2)	
2	Check level of lubricating oil in sump.	R		Lub^n Op^r		S.E.A. 20
	Weekly – "B" Service					
3	Clean oil bath air filter and top up with oil to the marked level.	S/D (items 3–7)	45	Lub^n Op^r	30 mins (items 3–6)	½ Litre S.E.A. 20
4	Drain fuel pump chamber or drain pot.		21	Lub^n Op^r		
5	Examine and, if necessary, oil the starting handle pawl and the portion of the shaft on which it fits.			Lub^n Op^r		
6	Lubricate the links and connections of all external controls.			Lub^n Op^r		
7	Check the tension of belts.		23	Mech Fitter	5 mins	
	Two Monthly – "C" Service					
8	Inspect fuel filter elements. Renew if necessary.	S/D (items 8–12)	47	Lub^n Op^r	1½ hrs (items 8–11)	2 off No 62127
9	Renew lubricating oil filter elements.		48	Lub^n Op^r		2 off No 36122
10	Drain lubricating oil sump, clean strainer, and refill with new oil.		53	Lub^n Op^r		10 Litres S.E.A. 20
11	Clean lubricating oil filler gauze.		51	Lub^n Op^r		
12	Grind in the valves.		66	Mech Fitter	8 hrs	

	Four Monthly – "D" Service					
13	Remove injectors, clean injector filter and test spray. Replace without interference if spray is satisfactory.	S/D	60	Mech Fitter	2 hrs	
	Six Monthly – "E" Service					
14	Renew fuel filter elements.	S/D	47	Lub^n Op^r	15 mins	
15	If engine shows signs of loss of power, remove cylinder head and piston, examine and check wear with recommended maximum wear allowances. If engine is performing satisfactory DO NOT DISTURB.			Mech Fitter		
	Annually – "F" Service					
16	Remove cylinder heads, examine inlet and exhaust valves, decarbonize, grind in valves.	S/D	70	Mech Fitter		
17	Withdraw and clean pistons, check that rings are free and the wear is within the recommended tolerance.		73	Mech Fitter		
18	Check and adjust valve and pump fuel tappet clearances.		71	Mech Fitter		
19	Examine large end bearings and check crankpins for ovality and scoring.		72	Mech Fitter		
20	Check fuel pump operating gear and governor for signs of undue wear.		76	Mech Fitter		
21	Clean exhaust ports, pipes and silencer.		75	Mech Fitter		
	2 Yearly – "G" Service		80			
22	Examine main bearings and check crankshaft for ovality and scoring.	S/D		Mech Fitter		
23	Remove and examine lubricating oil pump.			Mech Fitter		
24	Flush out all fuel and lubricating oil pipes.			Lub^n Op^r		
25	Inspect cooling system for obstruction to the air flow, particularly in the area adjacent to the fan.			Mech Fitter		
26	Renew connecting rod bolts.			Mech Fitter		

* R . . . . . . Running Maintenance
S/D . . . . Shut-down Maintenance

2.14 Maintenance schedule for a diesel engine

In practice, a series of job specifications provide this means. Each task listed in the maintenance schedule should have its own separate document or documents specifying the work to be done. The depth of the specification will vary to meet the needs of each situation as it must take into account the type of facility to be maintained, the complexity of the task, the control to be exercised over the work and the depth and scope of the planning. The most elementary form of job specification is a straightforward extract from the maintenance schedule:

> 'Withdraw and clean pistons, check that rings are free and the wear is within the recommended tolerance'
>
> (Item No. 17 of the Maintenance Schedule shown on pages 48–49.)

In the absence of other guidance, that type of specification (instruction) is wide open to any interpretation. On the spot technical decisions made during the task and the extent to which it is carried out will be based upon the previous experience, personal judgement and inclination of the person doing it. No two persons will have the same approach, opinion, attitude, standard or thoroughness; consequently, the work could be tackled in as many different ways as there are persons capable of doing it, with as many different variations in the results.

This short form of specification may suffice for the most basic of schemes where each task consists of only a few simple operations, but where the number of operations in each task is numerous or complex, or the adherence to manufacturer's recommended procedures and tolerances are vital, then a more detailed specification is essential. This extra information may be provided in the form of a check list not only for guidance but to ensure that no task is omitted. Figure 2.15 illustrates a typical example of a check list type of job specification which indicates the various inspections that must be made on a goods vehicle in order to comply with statutory regulations.* Should further details of a particular item be required they can, in this instance, be obtained by referring to the tester's manual (Fig. 2.16). For most other equipment the manufacturer's service manual would be consulted, failing this the tradesman must turn to past experience, but this allows the tradesman considerable latitude in tackling the tasks (see Fig. 2.17).

The higher the degree of planning the greater must be the control over the work operations. To ensure that the work is carried out effectively a comprehensive job specification is needed – one that sets out a complete list of working instructions, detailing exactly how each task is to be completed, the sequence of work, the methods and tools to be used, the limits and tolerances to be worked to and the safety measures to be observed. Figure 2.18 refers to Item 17 of the maintenance schedule shown on page 49, but this time the task has been set out as a fully detailed job

*Goods Vehicles (Plating and Testing) Regulations 1971.

BP FLEETCARE OPERATOR:

INSPECTION LIST

VEHICLE TYPE	REGn. No.	FLEET No.	SPEEDO Rdg.	DATE in	DATE out	OK	
						REPAIR REQ	
VEHICLE SHOULD BE CLEAN BEFORE INSPECTION						RECTIFIED	

MOT No.	ITEM		CHECK	REMARKS	No.	
	EXTERIOR INSPECTIONS					
1	Position of Legal Plate		Presence, Security, Prominence		1	
2	Details of Legal Plate		Legibility, Correct for vehicle		2	
3					3	
4					4	
5	Smoke Emission	*	Density		5	
6	Road Wheels & Hubs		Fractures, Distortion, Security		6	
7	Size & Type of Tyres		Size, Ply, Mixing of Tyres		7	
8	Condition of Tyres		Damage, Tread Depth & Width, Walls		8	
9	Bumper Bars		Security, & Condition		9	
10	Spare Wheel Carrier		Security, Condition, Wheel Security		10	
11	Trailer Coupling	*	Security, Wear, Safety Device, Deform'n.		11	
12	Coupling on Trailer				12	
13					13	
14	Condition of Wings		Presence, Damage, Security, Fouling		14	
15	Cab Mountings	*	Security, Condition, Locking Devices		15	
16	Cab Doors	*	Presence, Condition, Security, Operation		16	
17	Cab Floor & Steps	*	Security, Condition		17	
18	Driving Seat	*	Security, Condition, Adjustment		18	
19	Security of Body		Displacement, Security		19	
20	Condition of Body		Overall Condition, Safety, Security, Leaks		20	
21					21	
	INSIDE CAB INSPECTIONS					
22	Mirrors	*	Presence, Condition, Posn., Security		22	
23	View to Front	*	Obstruction		23	
24	Condition of Glass	*	Visibility, Cracks		24	
25	Windscreen Wipers	*	Presence, Function, Condn., Wiping Area		25	
26	Speedometer	*	Presence, Drivers View, Function, Illum.		26	
27	Audible Warning	*	Presence, Control Posn., Secty., Function		27	
28	Driving Controls	*	Completeness, Condition, Posn., Obstructn		28	
29					29	
30	Play at Steering Wheel	*	Not more than one fifth of diameter		30	
31	Steering Wheel	*	Security to Shaft, Condition		31	
32	Steering Column	*	End Float, Side Play, Flex Coupling		32	
33					33	
34	Air/Vacuum Warning	*	Presence, Visibility, Operation, Reserve		34	
35	Build-up of Air/Vacuum	*	Time required to operate warning		35	
36	Mech. Brake Hand Levers	*	Condn., Travel, Obstruction, Hold on.		36	
37	Service Brake Pedal	*	Condn. Secty, Travel Obstruction, Antislip		37	
38	Service Brake Operation	*	Leaks, Servo, Operation		38	
39	Air/Vac Hand Controls	*	Secty., Condition, Travel, Leakage		39	
40					40	

*NOT APPLICABLE TO TRAILERS

BP/210/70

2.15 A check list type of specification

5 Smoke emission

Method of Inspection

Check the density and colour of the emission from the exhaust outlet visually.

1. In the case of diesel engines depress the accelerator pedal firmly from the idling position to the maximum fuel delivery position. Immediately maximum engine speed is reached and the governor operates release the pedal until the engine slows to a steady idling speed. Ignore smoke emission from this FIRST acceleration. Repeat the operation to a maximum of four times until, if smoke is produced at all, the emission is considered to be of equal density for two successive accelerations.
2. In the case of petrol engined vehicles, speed the engine up intermittently to near full speed.

Note 1. This test is to be carried out with the vehicle as presented. There is no requirement that it must be carried out with the engine cold.
Note 2. When carrying out this test the engine must not be 'blipped'.
Note 3. This test will determine whether a vehicle is emitting excessive smoke as presented. If a vehicle satisfies the test it does not mean that it will not smoke excessively under different operating conditions.

Reasons for Rejection

The exhaust emission is coloured black haze or darker for two successive accelerations not including the first.

Motor Vehicles
Part 1 No 2

6 Road wheels and hubs

Method of Inspection

1. Make an inspection of each part of the road wheel, paying particular attention to whether there is:

(a) Any fracture on flanges.
(b) Any fracture on tyre retaining rings.
(c) Any welding breaking away.
(d) Any stud hole badly worn.
(e) Any wheel stud missing.
(f) Any wheel nut missing.
(g) Any wheel nut loose.
(h) Any tyre retaining ring, the ends of which are butting.
(i) Any wheel badly distorted.

2. Examine half shaft bolts, nuts and studs for security.

Note 1. It is not always possible to see the complete roadwheels on a vehicle from ground level, especially on twin wheels and on some vehicles where the body shrouds part of the wheels. In such cases the vehicle should be moved to expose the hidden parts of wheels or the examination should be completed from a position under the vehicle.
Note 2. The spare wheel is not included in the inspection, but if any defect on a spare wheel carried on the vehicle is seen, advise the driver.

Reasons for Rejection

1. Any fracture on wheels, except at the bridge over the valve.
2. Any welding breaking away.
3. Any sign of elongation of a stud hole on a wheel. ◀
4. Any wheel stud missing.
5. Any wheel nut missing.
6. Any wheel nut on a wheel loose. ◀
7. Any tyre retaining ring butting to such an extent that the ring is visibly displaced from its seating. (***Note:*** – On certain wheels, butting with slight displacement is acceptable.)
8. Any wheel so badly distorted that it is not running reasonably true.
9. Any half shaft bolt, nut, or stud loose or missing.

Motor Vehicles
Part 1 No 8(a) & (c)
Trailers
Part II No 5(a) & (c)

specification. Figure 2.19 is an example of a standard job specification which could be applied to all steam stop valves in the factory. It is this detailed type of specification that is usually the least acceptable to the tradesmen, who regard it as an affront to their craftsmanship and technical ability.

Little do they realize or appreciate the protection it affords them. Provided the work is carried out in accordance with the specification, they cannot be blamed if the maintenance is ineffective.

The employment of incentive schemes or work measurement makes the formulation and publication of a full job specification essential. The work content and method must be defined clearly so that all parties are aware of the basis upon which time allocation or payment is evaluated.

Thus, the object of a job specification is to ensure that:

1. The task or job is carried out in the manner intended.
2. The possibility of an operation being omitted is minimized.
3. Acceptable limits of wear, etc., and tolerances are clearly defined.
4. The tradesman or operator knows the work and how it is to be done.
5. The operation is always carried out in the same manner and the same allowances are always applied. Checking for wear, etc., is carried out in the same position and in the same way. Results over the course of time are all comparable. (Standardization.)
6. All persons doing the work, even for the first time, follow the same procedure. (Continuity.)
7. A reference standard is available with which the job and work content can be compared to detect any variation.

It is necessary to prepare a job specification for each of the tasks in the maintenance schedule. The specification should deal with one task only and detail exactly the inspection, servicing, adjustment, replacement, etc., required, together with instructions on the procedure to be adopted.

The specification should comprise an individual sheet or card for each task listed in the maintenance schedule. In cases where the task involves several different trades, i.e. mechanical fitter, electrician, instrument fitter, the trade responsible for carrying out a particular operation should be indicated clearly. Alternatively, a separate job specification can be made out for each trade.

To summarize

Each specification should indicate:

1. The identification number and name of the item of equipment.
2. The location of the item.
3. The maintenance schedule reference number, (and task reference) from which the task was extracted.

5000 MILE INSPECTION

WORKS/DEPOT________ MAKE ________ FLEET No________ DATE________

SPEEDO READING ________ LAST INSPECTION DATE ________ MILEAGE SINCE LAST INSPECTION________

CODING: IN ORDER ✓ REPAIRS REQUIRED ○ IMMEDIATE REPAIRS ×

STEERING

Check
- ☐ Play at steering wheel.
- ☐ Security of column and box.
- ☐ Clearance of steering arms throughout travel.
- ☐ Security of drop arm and method of locking.
- ☐ Wear of steering joints.
- ☐ Security of steering arms and method of locking.
- ☐ Wear in swivel pins and thrust races.
- ☐ Adjustment of lock stops.
- ☐ Wheel alignment.

BRAKES

Check
- ☐ If brakes operate correctly.
- ☐ Braking system for leaks with footbrake applied.
- ☐ Build up of vacuum or air pressure satisfactory.
- ☐ Brake warning gauge and/or device working.
- ☐ Security of brake pipes, servos or reservoir.
- ☐ Wear in connections.
- ☐ Security of clevis pins.
- ☐ Wear in pedal pivot bush.
- ☐ Wear in handbrake pivot bush, ratchet and pawl.
- ☐ Wear in cross-shafts.
- ☐ Wear on brake liners.
- ☐ Flexible pipes – wear, chafing or weeping.
- ☐ Handbrake cable fraying.
- ☐ Hydraulic reservoir fluid level.
- ☐ Correct adjustment foot and handbrake.

TRANSMISSION

Check
- ☐ Gearbox mounting.
- ☐ Clutch adjustment, clutch cross-shaft, clevis pins.
- ☐ Gearbox coupling flange and flange bolts.
- ☐ Speedometer drive and cable.
- ☐ Gear change mechanism and linkage.
- ☐ Propeller shaft, universal joints.
- ☐ Propeller shaft intermediate bearings and mountings.
- ☐ Differential housing, cover plate and oil filler bolts for security and oil seals.
- ☐ Differential drive oil seals.
- ☐ Drive shaft studs.
- ☐ Drive shaft oil seals.
- ☐ Differential lock operating pipes.

ELECTRICAL

Check
- ☐ Battery secure and terminals clean.
- ☐ Security of terminals and wiring clips.
- ☐ Headlamps, sidelamps, rear lights, stoplights, spotlights, indicator lights, horn and wipers.
- ☐ Charging rate. Lamp focus switches.

FRAME

Check
- ☐ Freedom from cracks, visibly true.
- ☐ Security of rivets and bolts.
- ☐ Security of all frame brackets.
- ☐ Serviceability of registration plates and reflectors.

WHEELS AND TYRES

Check
- [ ] Wheel nuts and studs.
- [ ] Wheels for cracks or broken flanges.
- [ ] Wheel bearings for wear.
- [ ] Tyres for cuts and irregular wear.

EXHAUST

Check
- [ ] Security and leaks.
- [ ] Density of exhaust smoke.

ENGINE

Check
- [ ] Cooling system, hose pipes, clips and fan.
- [ ] Belt adjustment, radiator cap, overflow, pipe, water circulation, water pump, radiator mountings.
- [ ] Engine mountings, inlet and exhaust manifolds, exhaust pipe security.
- [ ] Fuel pump mounting, fuel pump drive.
- [ ] Fuel pipes, injector pipes, return pipes, filters.
- [ ] Accelerator controls, venturi pipes, stop/start control.
- [ ] Starter and dynamo mounting.
- [ ] Tappets.
- [ ] Brake compressor or exhauster.
- [ ] Power steering pump, oil reservoir, steering oil pipes.
- [ ] Engine performance for excess fuel consumption, pump timing, defective injectors.
- [ ] Engine oil, oil filters and air filters.

CAB AND BODY

Check
- [ ] Mounting brackets and pads.
- [ ] Soundness of cab structure.
- [ ] Security of doors, locks, windscreens, windows and lights.
- [ ] Security of driver's seat.
- [ ] Soundness of cab floor boards.
- [ ] Fitting and serviceability of mirrors.
- [ ] Soundness and security of mudguards.
- [ ] Security of body mounting.
- [ ] Tipper ram mounting, hinge bar and brackets, tip gear.
- [ ] Oil reservoir.
- [ ] Ram seals.
- [ ] Neuter control valve.
- [ ] Pump engagement.
- [ ] Power take-off.
- [ ] Drive shaft.
- [ ] Outrigger bearing.
- [ ] Drive belts and pulleys.
- [ ] Air filter.
- [ ] Pipework and valves.

SPRINGS

Check
- [ ] Security of axle bolts, lock-nuts and spring clips.
- [ ] Broken or displaced leaves.
- [ ] Wear in shackles, pins and bushes.
- [ ] Security of hangers to chassis frame.
- [ ] Balance beam trunnions for wear.

M/M M/M M/M M/M

M/M M/M

TYRE TREAD DEPTH M/M

M/M M/M

M/M M/M M/M M/M

REPORT DEFECTS OVERLEAF WORK CHECKED & CLEARED

2.17 An example of a job specification for the periodic maintenance of a motor vehicle

JOB SPECIFICATION No 73

'Withdraw and clean pistons, check that rings are free and the wear is within the recommended tolerances.'

TO WITHDRAW PISTON

1. Remove cylinder heads . . . . Job Specification No 70
2. Uncouple connecting rod big end and bearing. Job Specification No 79
3. Lift piston and connecting rod from cylinder.
4. Remove gudgeon pin.

   To assist removal immerse piston assembly in hot oil, or stand piston on hot plate.

TO REMOVE PISTON RINGS

5. Soak piston assembly in paraffin to soften carbon deposits.
6. Spring open rings and insert thin metal strips between the rings and piston at four different points.
7. Remove rings by sliding them over the metal strips.
8. Thoroughly clean pistons and rings. Flush through oil holes and passages with syringe using paraffin.

CHECK DIMENSIONS OF ALL RINGS

9. Check gap. (Pressure and Scraper rings similar.)

   Each ring must be checked and measured in the same cylinder bore from which it was removed.
   In the case of new rings measurements must be made in the cylinder the ring will eventually occupy.
   Check gap at unworn portion of the liner.
   Insert ring squarely in cylinder bore.
   Use Jig No 1052 to locate.

10. Measure gap with feelers.

Designed gap size	0·25 mm to 0·38 mm 0·010" to 0·015"
Renew ring when gap reaches	0·76 mm 0·03"

11. Replace each ring back into its original piston groove.

    Use the thin metal strips to slide rings over piston.

12. Check clearance between piston ring groove and piston ring edge.

Designed clearance	0·025 mm to 0·073 mm 0·001" to 0·003"
Renew ring when clearance exceeds	0·2 mm 0·008"

NOTE: If designed clearance cannot be obtained with new standard ring, open out groove and fit over-width ring.

CHECK GUDGEON PIN AND SMALL END BUSH

13.

Designed clearance	0·04 mm to 0·064 mm 0·0015" to 0·0025"
Renew when clearance exceeds	0·1 mm 0·004"

REFIT GUDGEON PIN AND CONNECTING ROD TO PISTON

14. Immerse piston in hot oil or stand on hot plate to ease insertion of pin into piston.

REPLACE PISTON

15. Smear lubricating oil on piston piston and liner.
16. Stagger piston ring gaps.
17. Using Jig No 1056 to hold in piston rings, slide piston into liner.
18. Couple connecting rod big end and bearing. Job Specification No 79
19. Replace cylinder heads. Job Specification No 70.

2.18 Job specification No. 73

4. The job specification reference number. (Each job specification should be allocated its own individual reference number.)
5. The frequency at which the particular task must be carried out.
6. The trade or trades required to carry out each operation.
7. The specific details of the work to be done – the procedure and method, details of inspections, tests, servicing, etc., allowable limits and tolerances.
8. Components to be replaced.
9. Special tools and equipment to be used.
10. Reference drawings, manuals, etc., applicable.
11. Safety procedures to be followed.

The specification issued to tradesmen should also include a footnote or a standing order to the effect:

> 'Any minor defect observed in the course of the specified work but not listed for attention should, if possible, be corrected *and* reported. If it is not possible or convenient to correct the defect in the course of the specified work it should be reported. If the defect is considered sufficiently serious it should be brought immediately to the attention of the authorities.'

2.7. The maintenance programme

The maintenance programme is a list allocating specific maintenance to a specific period.

Having established *what* is to be maintained, and *how* it is to be maintained, we must now consider *when* it is to be maintained.

A programme to indicate when each facility shall receive its scheduled maintenance must be written. Its main purposes are:

1. To set out a plan of work.
   - (a) To spread the maintenance work load evenly over the year.
   - (b) To ensure that no facility or maintenance task is omitted.
   - (c) To ensure that the required maintenance is carried out at the specified frequency.
   - (d) To co-ordinate the maintenance of associated facilities.
   - (e) To co-ordinate maintenance with production requirements.
2. To present an overall picture of maintenance work, present and future commitments (on short and long term basis).
   - (a) To assist forward planning, ordering of spares, future labour requirements, basis for budgetary control.
3. To act as a reminder of future maintenance events (on a short term basis).
   - (a) To formulate weekly work plans (for immediate future).
   - (b) To arrange for availability of production plant.
   - (c) To arrange or check availability of labour, spares, sub-contractors, etc.

The time span of the programme is flexible, usually it is on an annual

FACILITY IDENTIFICATION Nº	MAINTENANCE SCHEDULE REF Nº____ ITEM Nº____	JOB SPECIFICATION Nº STANDARD 69–1
FACILITY DESCRIPTION STEAM STOP VALVE	JOB DESCRIPTION:- OVERHAUL OF STEAM STOP VALVE	
LOCATION:- ALL STEAM LINES ON SITE	TRADE:- MECH FITTER	FREQUENCY:- ANNUALLY

1. OBTAIN Permit-to-Work.
2. Shut off the two Steam Stop Valves on each side of valve to be overhauled. Secure in the closed position with lock and chain.
3. Allow time for line steam pressure to drop to zero.
4. Remove valve from line. Fit blanks to open ends of line if no replacement valve is available.
5. Remove bridge cover nuts and withdraw cover, spindle, and valve lid.
6. Dismantle gland, gland packing, handwheel, spindle and valve lid.
7. Examine the following components.
   GLAND for bending and cracking.
   VALVE SPINDLE in way of packing and at screw thread.
   Check for truth.
   BRIDGE COVER screw thread for wear and with valve spindle for fit and movement.
   VALVE LID FACE for scoring and pitting, machine if necessary.
   STUDS on valve chest for condition and fit, hammer test for brittleness.
   VALVE SEAT FACE for scoring and pitting, machine if necessary.
   VALVE CHEST FLANGES for condition and flatness.
8. Compare dimensions of valve seat, valve lid, and spindle with manufacturers drawings.
   All spindles, valves, etc., to be brought back to drawing dimensions on renewal.
9. Water test valve chest to twice working pressure.
10. Assemble valve and water test to twice working pressure.
11. Remove blanks and replace valve in line.
12. Remove chains and locks on adjacent stop valves.
13. RETURN Permit-to-work.

DRAWINGS REQUIRED	
SPECIAL TOOLS REQUIRED	
OBSERVE SAFETY REGULATIONS	CLEAR SITE ON COMPLETION

2.19 A standard job specification (for the annual overhaul of steam stop valves)

basis, but where plants operate for two or even three years between major overhauls the planning period is often extended to cover these cycles.

The programme should be prepared in consultation with the production department which is able to advise on production schedules and plant availability so that maintenance fits in with production requirements. The various maintenance operations must be time-tabled so that they occur at the right frequency and within the recommended interval of time. Associated facilities should be programmed together – diesel engines and their coupled generators, air compressors and their connected receivers, electric motors and their associated switchgear, pumps and their driving motors. Certain facilities, in order to comply with the Factories Acts, must receive periodic attention. To assist the planner in this respect, an extract of the regulations is given in Appendix 2.

As the programme is built up the weekly work loads must be assessed (from the work times quoted in the relevant maintenance schedules) to ensure that, as far as possible, the total work load is evenly spread over the year. Initially, this may require that maintenance of certain facilities be brought forward. This could well apply to those facilities that must be periodically inspected or tested to comply with the requirements of the Factories Acts. Unless deliberately changed, the test dates of facilities installed at the same time could continue to coincide throughout their operational life. A large battery of pressure vessels all installed at the same time would normally all come up for re-test in the same period 26 months later, thus placing a heavy load on the maintenance department. But, by bringing forward some of the testing, perhaps in some cases by as much as 18 months, and staggering the remainder, the work load can be spread evenly over the 26 months.

As one of the aims of planned maintenance is to reduce downtime, the programme must ensure that the planned downtime for shut down maintenance is considerably less than the downtime previously experienced. Planned maintenance will not guarantee freedom from breakdowns so a time allowance must be included for such unexpected contingencies when assessing total downtime.

Initially, it is prudent and practical to include only a few items in the programme, but as experience is gained so the number of items covered can be gradually increased.

There are a number of ways of presenting the programme. The more usual methods and the ones commercially available involve planning charts (or boards) or Visible Record cards (see Chapter 5); some systems employ a combination of both. But whatever method is used, it must be capable of quick modification, as seldom is it possible to plan with any certainty on a long term basis.

Planning charts

The facilities to be maintenanced are listed down the left hand side of

LOCATION: NO. 2 SHOP

PLANT ITEM	PLANT NO.	PERIOD (MONTHS) MECH.	PERIOD (MONTHS) ELEC.	PERIOD (MONTHS) LUB.
CENTRE LATHES				
12" LATHE	01-08	6 12	6 12	12
8" LATHE	01-12	6 12	6 12	6
8" LATHE	01-14	3 6 12	6 12	6 12
14" LATHE	01-15	6 12	6 12	12
17" LATHE	01-19	6 12	6 12	12
6" LATHE	01-23	3 6 12	6 12	12
CAPSTAN LATHES				
SIZE 1 LATHE	02-03	3 12	6 12	12
SIZE 3 LATHE	02-05	3 12	6 12	12
SIZE 2 AUTO LATHE	02-06	1 6 12	6 12	3 12
SIZE 4 AUTO LATHE	02-09	1 6 12	6 12	3 12
DRILLS				
BENCH DRILL	05-03	6 12	1 12	Not Reqd
PEDESTAL DRILL	05-12	6 12	12	12
4-SPINDLE DRILL	05-17	6 12	6	6
MULTI-SPINDLE DRILL	05-29	6 12	6	3
RADIAL DRILL	05-34	6 12	6 12	12
RADIAL DRILL	05-36	6 12	6 12	12
MILLING MACHINES				
HORIZONTAL MILLER	09-04	3 6 12	6 12	6
VERTICAL MILLER	90-60	6 12	6 12	6
GRINDING MACHINES				
UNIVERSAL GRINDER	10-02	3 12	6	6
SURFACE GRINDER	10-03	6 12	6	6
MISCELLANEOUS				
BUFFING MACHINE	14-07	12	12	Not Reqd
GUILLOTINE	17-13	12	12	
CROPPER	17-15	3		6

WEEK NUMBER	1	2	3	4	5	6	7	8	9	10	11	12	13	14	15	16	17	18	19	20	21	22	23	24	25	26
MONTH	JANUARY				FEBRUARY				MARCH					APRIL				MAY					JUNE			
WEEK STARTING	4	11	18	25	1	8	15	22	1	8	15	22	29	5	12	19	26	3	10	17	24	31	7	14	21	28
CENTRE LATHES																										
12" LATHE	12																									6
8" LATHE								12																		
8" LATHE			3												12											
14" LATHE																						12				
17" LATHE					6																					
6" LATHE												6													3	
CAPSTAN LATHES																										
SIZE 1 LATHE		12												3												
SIZE 3 LATHE						3													12							
SIZE 2 AUTO LATHE			1				1				6				1					1				3		
SIZE 4 AUTO LATHE	3				1				1				12					1				1				3
DRILLS																										
BENCH DRILL				12				1				1					1				1				1	
PEDESTAL DRILL										6																
4-SPINDLE DRILL																										
MULTI-SPINDLE DRILL							3											6								
RADIAL DRILL									12											12						
RADIAL DRILL														12												
MILLING MACHINES																										
HORIZONTAL MILLER										3													6			
VERTICAL MILLER																								12		
GRINDING MACHINES																										
UNIVERSAL GRINDER				6													3									
SURFACE GRINDER																					6					
MISCELLANEOUS																										
BUFFING MACHINE																										
GUILLOTINE																										
CROPPER																										
																EASTER HOLIDAY										

WEEK NUMBER	27	28	29	30	31	32	33	34	35	36	37	38	39	40	41	42	43	44	45	46	47	48	49	50	51	52
MONTH	JULY				AUGUST					SEPTEMBER				OCTOBER				NOVEMBER					DECEMBER			
WEEK STARTING	5	12	19	26	2	9	16	23	30	6	13	20	27	4	11	18	25	1	8	15	22	29	6	13	20	27
CENTRE LATHES																										
12" LATHE																										
8" LATHE									6																	
8" LATHE		3														6										
14" LATHE																						6				
17" LATHE				12																						
6" LATHE													12												3	
CAPSTAN LATHES																										
SIZE 1 LATHE	6														3											
SIZE 3 LATHE							3												6							
SIZE 2 AUTO LATHE		1						1				12				1				1				3		
SIZE 4 AUTO LATHE				1						1				6				1				1				
DRILLS																										
BENCH DRILL			6						1				1				1				1				1	
PEDESTAL DRILL											12															
4-SPINDLE DRILL																		12								
MULTI-SPINDLE DRILL								3												6						
RADIAL DRILL										6																
RADIAL DRILL															6											
MILLING MACHINES																										
HORIZONTAL MILLER																							12			
VERTICAL MILLER																								6		
GRINDING MACHINES																										
UNIVERSAL GRINDER			12														3									
SURFACE GRINDER																					12					
MISCELLANEOUS																										
BUFFING MACHINE																			12							
GUILLOTINE	12																									
CROPPER																										
					ANNUAL SHUT-DOWN																					CHRISTMAS HOLIDAY

2.20 A maintenance programme

the chart or board. The time scale, usually in days or weeks, is marked along the top edge. By means of colour coded pins, pegs or crayons to represent the various maintenance operations, the work in progress and future commitments are depicted clearly.

Examples of this type of programming are shown in Figures 2.20 and 2.21, while 'peg-board charts' (page 151) describes the system in more detail.

In Fig. 2.20 'the plant items are listed on the left, with their reference numbers, and alongside these a table summarizes the mechanical, electrical and lubrication maintenance each requires at various intervals. This ready-reference summary is a valuable aid to programming and lessens the danger of operations being overlooked. The maintenance required by each machine is then entered in the appropriate week columns (with due allowance for holiday periods), the twofold object here being to ensure that every operation falls within the specified time limits and that the weekly maintenance load shall be as nearly as possible constant. Maintenance cycles are commonly based on multiples of three months, and it will therefore be convenient to work to a planning year of 48 working weeks.

Week No. . . . . . . W/E . . . . Department		Mechanical ☐ Red Electrical ☐ Blue Instrumentation ☐ Yellow Lubrication ☐ Green						
Iden. No	Facility	Sun	Mon	Tu	Wed	Thu	Fri	Sat

2.21 A layout for a weekly maintenance plan

For closer control it may in some cases be desirable to subdivide the weeks into five or (for continuously running plant) seven working days. An advantage of five-day programming is that it eliminates routine weekend overtime, while allowing a margin for possible maintenance backlogs'.[8]

Visible record cards

A separate sheet or card is allocated to each facility in the maintenance programme. The sheets are filed in books or trays so that they overlap as those used for the facility register described on page 31. The lower visible edge bears not only the identification number of the facility concerned but a printed section showing either months or weeks of the year. The date when maintenance is next due is indicated by placing a small coloured signal at the lower edge of the card over the appropriate month or week. Different coloured signals denote the various maintenance operations. By looking down the page or tray it is easy to align the various tasks to be done on a particular date (see Fig. 2.22).

As only the lower 9·5 mm ($\frac{3}{8}$ in) of each document is used for programming, there is sufficient space available for other purposes, usually as a history record card (see Fig. 2.27). Occasionally, it is the lower edge of the facility record card that is used for programming (see Fig. 2.5).

Frequently, the two systems, planning charts and visible record cards, can be combined with advantage – the cards indicating only the month maintenance is due, while charts give the detailed programme of all the operations to be carried out in that particular month. Large plants may wish to carry this arrangement still further, the cards indicating the week number, the charts a detailed programme for the week (see Fig. 2.21).

A rather different form of programming is provided by the Kalamazoo Strip Index, further details of which are given on page 152.

Occasionally, small jobs arise which occur only once and which had not been visualized.

Such jobs might result from:

Works suggestion schemes.
Small improvements to plant and equipment.
Latent defects that have become apparent during operation.
The correction of minor faults in newly installed plant.

Usually, these jobs do not require immediate attention and can be carried out at the earliest convenient shutdown, but they must be included in the programme, together with a system of periodic review to ensure that they are not overlooked.

There will also be occasions, other than breakdown or a planned maintenance stop, when the plant must be shut down. The opportunity should not be lost for either carrying out these small jobs or bringing forward some of the programmed work.

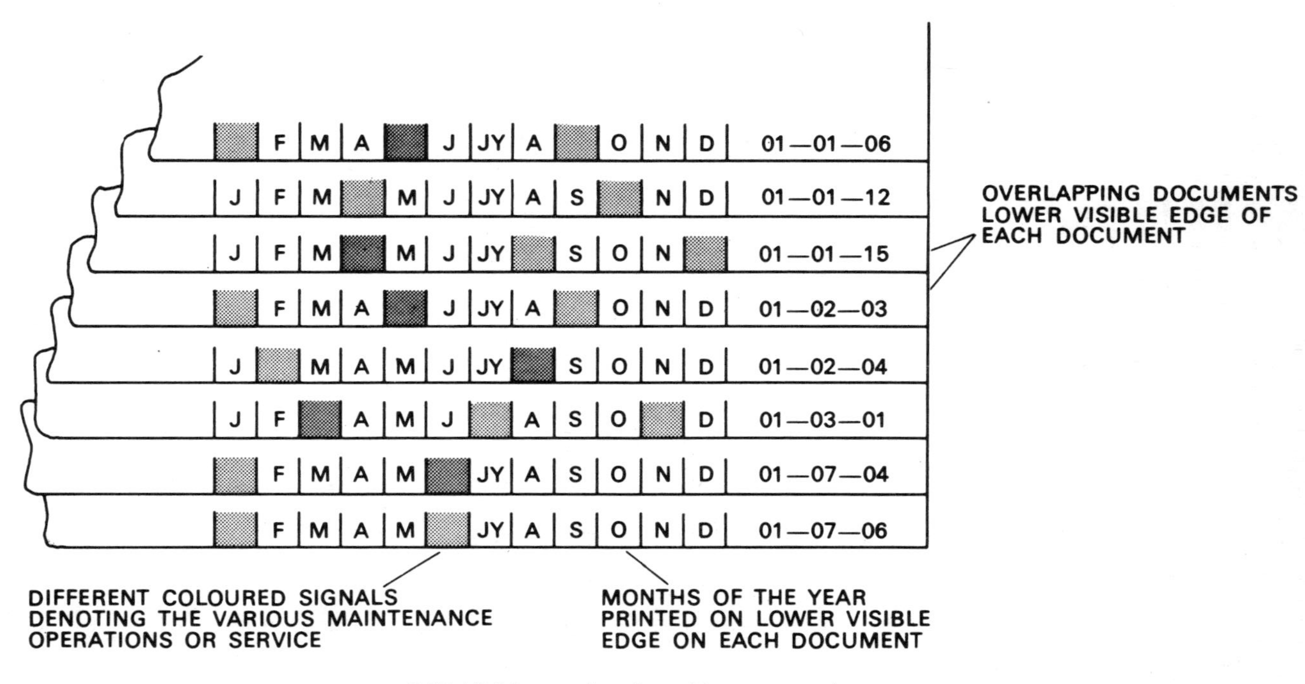

2.22 Visible record cards used for programming

Even with planned maintenance unexpected breakdowns will occur occasionally, but to withdraw labour from programmed work to carry out immediate repairs can have a disruptive effect upon the programme. When a fault is reported, it is not always necessary, or even in the best interests, to divert labour to deal with it immediately. If immediate repairs are not imperative, but can be fitted in with other work, the disruptive influence can be minimized. By rating faults and breakdowns to a priority scale, the maintenance department has an indication of their relative importance and can then plan their repair accordingly. An example of such a breakdown repair rating is:

A* – Must be repaired immediately.
(Faults which immediately affect the production process or create a safety hazard).
B* – Must be repaired within 1 day.
C – Must be repaired within 3 days.
D – Must be repaired within 1 week.
E – Can be carried out at the first available opportunity.

Initially, persons reporting breakdowns under this system tend to quote a higher rating than is warranted. When it is found that the maintenance department can keep to its obligations and that the system improves the service, the confidence generated will lead to more realistic reporting.

2.8. The control cycle

The control of a planned maintenance system depends upon monitoring the results and in the light of these implementing, if needed, corrective action. Automatic process control is a continuous cycle of:

Sampling the output of the plant.
Analysing the sample.
Applying correction to the plant, if necessary.

The control system must be capable of accommodating and responding quickly to continually changing conditions. A system that cannot do this will soon fall into disuse. This requirement as well as a control loop (Fig. 2.23) are equally applicable to planned maintenance systems but in the latter case the corresponding instruments of control are:

Sampling the effect of the maintenance – job report.
Analysing the effect of the maintenance – history card.
Applying corrective action – Maintenance schedule, Job Specification, Maintenance programme } revised as necessary
(see Fig. 2.24).

*Note: Owing to the urgency of type A and B breakdowns, it would be sufficient to report them by a phone call – the official reporting procedure being by-passed and the paper work completed later.

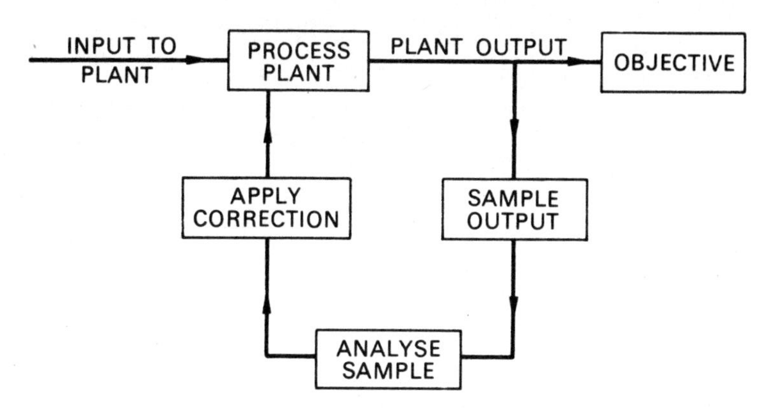

2.23 A simple process control loop

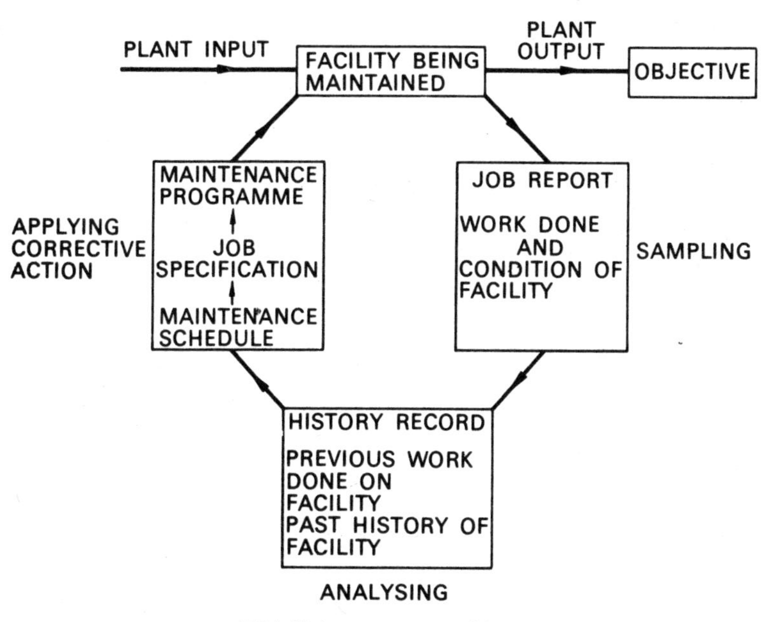

2.24 Maintenance control loop

2.9. Job report

The job report is a statement recording the work done and the condition of the facility.

For any planned maintenance scheme to be effective there must be a continuous flow of information to *and* from the person(s) doing the work. This feed-back is essential for the control and adjustment of the plan. The planner must know exactly what is happening – unless he knows that something is wrong or requires attention he cannot initiate remedial action. Every system must provide the person who has made the inspection or carried out the work the opportunity to report back his findings on the condition of the facility, the work done, and the work still to be done. Verbal reports are not reliable since they become forgotten or distorted. A document, a job report (Fig. 2.25), on which the relevant information can be recorded is issued to, and must be completed by, the person as and when he does the work.

Job Report	Date	Report No
Name of Tradesman	Clock No	Trade

Details of Report

Item:

Defect/condition:

Result:

Corrective action:

Spares/materials used:

Measurements/observations

Remarks:

Time taken:

Facility	Location	Identification No

2.25 A job report

The report should include information concerning:

Work carried out.
Corrective action taken.
Defects found, corrected, and their cause.
Defects observed but not corrected.
Work detailed on the job specification but not carried out or completed.
Major components replaced.
Measurements, of clearances, wear etc., taken.
Time taken to complete the work.
General observations, condition of the facility.

The job specification illustrated in Figs. 2.15 and 2.17 also doubles as a job report, the results of the inspection are recorded within the body of the specification. Against each of the items basic information is written in code in the right hand column. For example:

✓ indicates In order
O indicates Repairs needed
✓Ø indicates Rectified.

This can be supplemented by written details in the remarks column. Often the reverse side of the job specification is utilized as a report form, but if the same specification cards are to be used repeatedly then separate report forms are necessary.

To most tradesmen, clerical work and report writing are alien tasks. The facts and object of a report can be lost or misconstrued, simply because the writer could not adequately express himself on paper. In addition, conditions in the workshop and on site are not always conducive to lengthy or accurate reports. Consequently, the job report should be so designed that, although it supplies all the necessary information:

(a) it can be completed easily with the minimum of mental effort,
(b) the need for writing is kept to a minimum (perhaps by the use of code symbols), and
(c) it is self indicative of any work or item that has been omitted (see, for example, Fig. 2.26).

Occasionally, a tradesman is called upon to correct a minor defect – a blown fuse, a sheared pin, a defective warning light. In isolation, it may appear trivial and not worth the bother of making out a job report, indeed the paper work may take longer to complete than the actual job. But if the same defect occurs several times it could be the signal of more serious trouble on the way. It is only when these individual incidents are pieced together and viewed as a whole that their true meaning becomes apparent. Thus, seemingly unimportant faults should be reported so that they can be entered in the history card and so help to build up the complete maintenance picture.

If and when sub-contractors are used to carry out maintenance work scheduled in the programme, it is most important that they submit job

FACILITY IDENTIFICATION N°	MAINTENANCE SCHEDULE REF N° ITEM N°	JOB SPECIFICATION N° STANDARD 87–3
FACILITY DESCRIPTION PIPE-LINE BENDS	JOB DESCRIPTION CHECK PIPE-LINE BENDS FOR WALL THICKNESS	
LOCATION: ALL PIPE LINES ON SITE	TRADE PIPE FITTER	FREQUENCY: ANNUALLY

PIPE LINE N° . . . **BEND N° . . .**

APPLY D-METER AT THE NUMBERED TEST POSITIONS AND RECORD THE READINGS OBTAINED IN THE APPROPRIATE PANELS ON THIS REPORT FORM

NOTE:- THE TEST POSITIONS ARE INDICATED BY YELLOW PAINT CIRCLES 25 mm dia.
THE POSITION NUMBERS ARE DESIGNATED BY BLACK NUMERALS PAINTED IN THE MIDDLE OF THE CIRCLES.

POSITION N°	READING mm
1	
2	
3	
4	
5	
6	

DRAWINGS REQUIRED	DATE
SPECIAL TOOLS REQUIRED D-METER	SIGNED

OBSERVE SAFETY REGULATIONS	CLEAR SITE ON COMPLETION

2.26 A job specification/report. Note: any reading omitted is self indicative

reports, but the depth and detail of the reports would have to be agreed before the start.

In accordance with the Factories Acts, certain facilities – pressure vessels, boilers, lifting gear, etc. – must be periodically examined by a competent person, usually an insurance surveyor. The report of all such examinations must be made on an approved form and kept in the general register, but relevant extracts of the report can be used with advantage as a normal job report for entry into the planned maintenance history card.

2.10. The history record

The history record is a document on which information about all work done on and/or by a particular facility is recorded.

Planned maintenance is only a means to an end. It is a system or procedure for controlling the maintenance activities towards an approved objective.

No system is perfect. Plans, however carefully conceived and conscientiously followed, do not always work out as anticipated. The unexpected will occur, conditions will change, but if the system is to be a viable proposition regular checks are necessary to ensure that:

1. the assumptions upon which it is based are valid, and
2. it fulfils its intended purpose.

Item (1) can be considered from two organizational or management levels:

(a) Company or works management level.

The continuing validity of the particular system within the overall company or production policy.

(b) Maintenance management level.

The validity of the technical assumptions made within the system.

Many of the assumption made in (b) may depend upon policy decisions taken by higher management and involve issues that are beyond the immediate control of the maintenance department. (1) and (2) will be considered purely in relation to the planned maintenance system as applied at maintenance management level.

The success of any system can be evaluated only from the results it achieves, it is upon this evidence that decisions are made for future action. If the system is to be effective as a controlling medium, it must monitor its progress to check that it is producing the desired results, and assess, if necessary, the form and extent of corrective measures needed. Thus, arrangements for checking regularly the effect of the maintenance activity must form a part of the system.

A simple monitoring/control cycle for planned maintenance is illustrated on page 64. Maintenance information is fed into and is stored by the history record card. The subsequent analysis and interpretation of the recorded data assesses the effectiveness of the maintenance function and highlights any shortcomings in the system or its application. It is on this evidence that the most appropriate action can then be initiated. But if the assessment is to be of value, adequate, up-to-date records and information are necessary.

Each facility or item contained in the plan has an individual record which chronicles the maintenance events during its operational life. We must decide:

(a) what information is to be recorded, and
(b) how the information is to be recorded and stored.

Apart from basic details of name, identification number, location, etc., the precise range of information to be recorded will depend upon the type of facility concerned. Initially, it may be difficult to decide what information is necessary and the tendency might be to record everything just in case, but with experience the key factors emerge and the information can be rationalized and summarized effectively. It is pointless recording masses of information if it cannot be properly used or understood – less information of an elementary nature is more readily assimilated and appreciated.

For most purposes it is sufficient to log, in chronological order, such details as:

Inspections, repairs, servicing and adjustments carried out.
Breakdowns and failures, their results, their causes, corrective action taken.
Work done on the facility, components repaired or replaced.
Conditions of wear, tear, errosion, corrosion, etc.
Measurements or readings taken, clearances, results of tests and inspections.
The time and cost to carry out the maintenance or repair.

Although most of this information will be taken from job reports, some might be obtained through other departments – stores requisitions, plant logs, time and cost sheets.

There are many different systems for recording and storing information – some have been described elsewhere in the book, but whatever method is used the presentation should be simple to enable the pattern of events to be visualized clearly, in the right perspective and, where applicable, facilitate 'before and after' comparisons. To provide continuity the record document should have sufficient space to record the information of several years.

Typical examples of history record cards are illustrated in Figs. 2.27 and 5.6, the latter also combines the functions of job specification/job report.

Pictorial representations are usually more effective than lists of tabulated figures or written data. Frequently, the history record can, with advantage, be supplemented by graphical records (Fig. 2.28). Such graphs, or charts, can show at a glance a particular behavioural trend or

HISTORY RECORD CARD		Date from . . . . Date to	Sheet No	
			Cost/Time	
Date	Job Report No	SUMMARY OF: Item — Defect — Cause — Corrective Action — Spares/Materials Used	Planned Maintenance	Unplanned Maintenance

J	F	M	A	M	JU	JY	A	S	O	N	D	Facility	Location	Identification No

2.27 A history record card

REPLACEABLE GRAPH CARD.
OLD CARD IS FILED AND IS REPLACED BY A NEW CARD WHEN A REPLACEMENT BEND IS INSTALLED IN THE PIPELINE BUT THE MAIN HISTORY RECORD CARD IS CONTINUOUS.

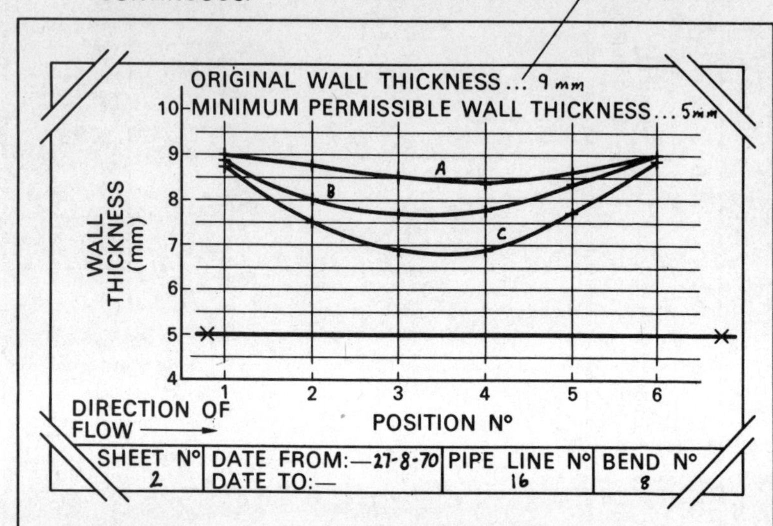

DATE	SHEET Nº	CURVE	POSITION Nº AND READING						REMARKS
			1	2	3	4	5	6	
12-6-66	1	A	9	8·5	8·3	8·3	8·6	8·9	
18-7-67		B	8·9	7·8	7·7	7·6	8·2	8·8	
5-7-68		C	8·8	7·6	6·9	6·8	7·6	8·6	
10-7-69		D	8·7	7·2	6·5	6·3	7·3	8·4	
15-7-70		E	8·6	6·4	6·7	5·2	6·5	8·2	RENEW BEND
20-7-71	2	A	9	8·7	8·5	8·4	8·6	9·0	NEW BEND 27-8-70
10-9-72		B	8·9	8·0	7·7	7·8	8·4	8·8	
4-9-73		C	8·8	7·5	6·9	6·9	7·7	8·7	

J	F	M	A	M	J	J	A	S	O	N	D	MINIMUM PERMISSIBLE WALL THICKNESS. 5mm	PIPE LINE Nº 16	BEND Nº 8

2.28 A history record card. This card should be used in conjunction with the job specification/report shown in Fig. 2.26

the rate at which a particular function is changing, and so forecast failure before any other indication is apparent (see Fig. 2.29). As behavioural patterns become familiar the various events can be anticipated with greater accuracy.

Other functions that can be usefully plotted include:

Resistance across filters, heat-exchangers, etc. (Fig. 2.30).
Temperature-differentials.
Fall in output of engines, generators, pumps, compressors, etc. (Fig. 2.31).
Rise in power input or fuel consumption.

If the information is handled by an automatic data processing system, it is necessary to convert data relating to maintenance events and faults into an acceptable, usually numerical, code. The details of the code are tailored to suit the circumstances, but in principle it can be constructed along similar lines to those described for plant identification (section 2.3).

The information is converted into a code consisting of a series of digits. Each digit, or group of digits, representing a particular aspect of the maintenance function or fault. The following is a very simple example to illustrate the application:

The first digit or group represents the field of activity involved.

1 – Mechanical	4 – Lubrication
2 – Electrical	5 – Instrumentation
3 – Electronic	6 – Hydraulic

The second digit or group represents the form of maintenance.

1 – Planned preventive running maintenance
2 – Planned preventive shut-down maintenance
3 – Planned corrective shut-down maintenance
4 – Emergency maintenance.

The third digit or group represents the type or the specific component involved.

The fourth digit or group represents the type of fault, failure; service or adjustment carried out.

	Field of activity	Form of maintenance	Component	Adjustment carried out
	↓	↓	↓	↓
The code →	1	3	7	8
	↓	↓	↓	↓
Translated into →	Mechanical maintenance	Planned corrective shut-down	Stirrer shaft gland	Repacked

The information is entered on to the record card usually by a clerk who, although quite capable of carrying out the clerical duties involved, does

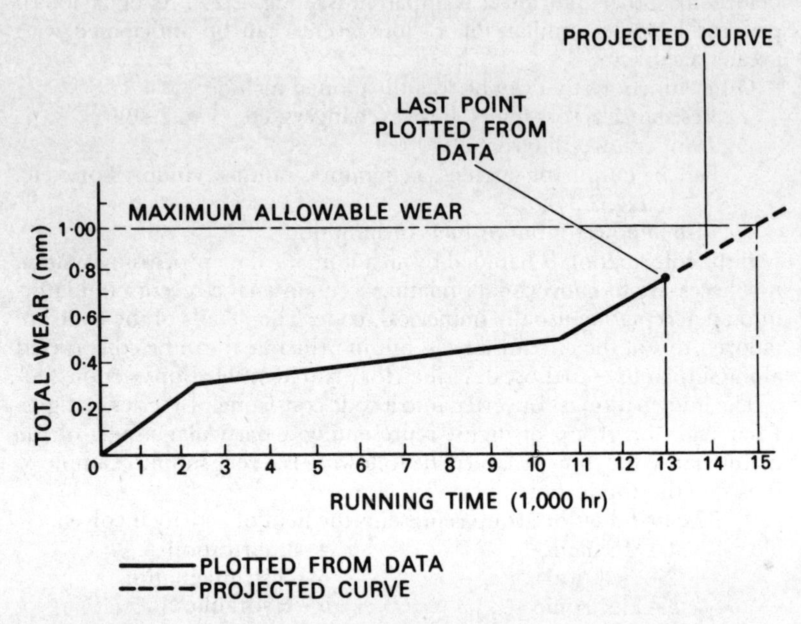

2.29 The projected curve suggests that the component has another 2000 running hours before reaching its maximum allowable amount of wear. This knowledge enables its replacement to be programmed well in advance

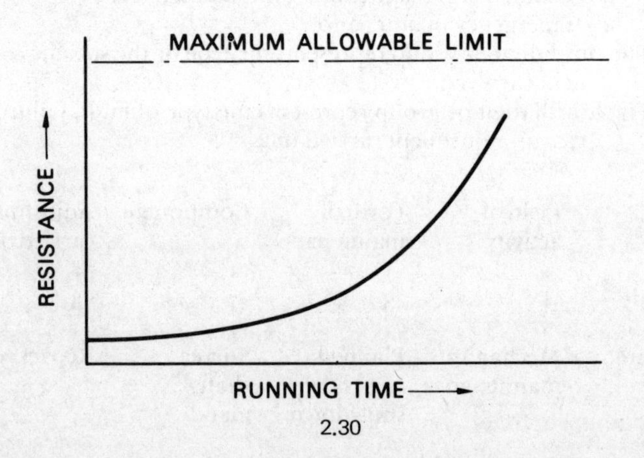

2.30

not have the technical knowledge to understand the implications or importance of the entries. There is nothing wrong with a clerk making these entries and keeping the records provided that they are scrutinized regularly by a qualified person who, from his interpretation of the presented facts, is able to initiate appropriate action. There is no point in recording the information if it is not usefully employed.

As the recorded information accumulates, so the effects of the planning can be visualized and studied:

Is the objective being achieved?
Is the plan adequate?
Is the maintenance adequate?
Are the facilities over or under maintained?
Is the form of maintenance correct?
Are the maintenance frequencies correct?
What is the breakdown frequency of a particular item or component?
What is the maintenance cost of a particular unit?

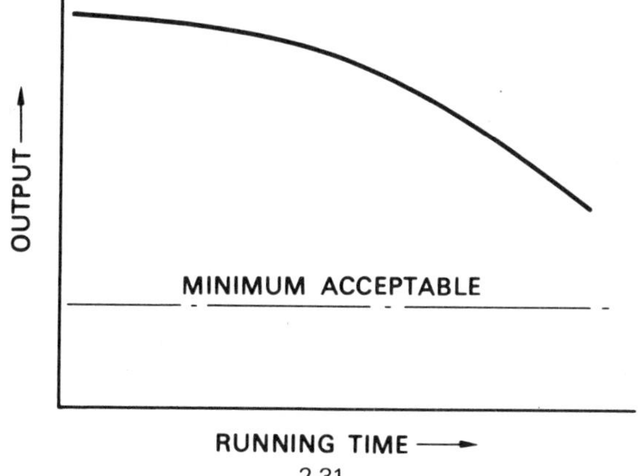

2.31

By analysing the various factors we can highlight the causes of failure and the weaknesses in the plan. As these are corrected and the appropriate modifications incorporated into the maintenance schedules, job specifications and maintenance programme, breakdowns should decrease and the amount of unplanned work gradually diminish. When a component fails regularly, and its failure cannot be prevented, then a new part must be fitted just before the old one, according to the records, is due to fail. The absence of breakdowns, or regular inspections and servicing finding nothing to report, is not necessarily a sign of successful planning – it could be over-maintenance, in which case the intervals between maintenance could safely be extended until the correct balance is reached.

There comes a time in the life of most equipment when we must decide whether it is more economical to purchase a new machine or to continue maintaining the existing one. As the equipment grows older so its need for maintenance increases. In this respect, maintenance is purchasing extra production capacity or time, but it is uneconomic if the cost of the maintenance is greater than the profit from the extra production. It is from the evidence contained in the history record that the economics of replacement can be realistically evaluated.

3

Auxiliary Components of the System

3.1. Designing out maintenance

A machine or installation that is technically sound, produces the required results and can be manufactured at a specified cost, is the aim of a designer. While every effort is made to achieve this objective, the 'built in' maintenance incurred is too often overlooked. Often, insufficient attention is given to those details that reduce the need for maintenance or facilitate unavoidable maintenance. Maintenance is a recurring expense that must be taken into account by the purchaser and should be considered by the designer. Many maintenance needs can be eliminated at source by 'designing out' their causes at the drawing board stage, but the designer must have operational information fed back to him in order to assess the maintainability of his designs or improve future models. It is the user, in this instance the plant or maintenance engineer, with his operational experience including his maintenance records, who is best qualified to provide the factual information (the failures and their causes, the nature and frequency of the maintenance, the difficulties experienced in carrying out this work, the replacement parts fitted, etc.) upon which ideas for improvements can be based. But for these ideas to become a practical reality there must be close co-operation between designer, manufacturer and user.

Although this policy is applied most effectively at the design stage, the opportunity to incorporate improvements during installation, at overhauls or when new ideas present themselves should be taken.

Maintenance problems are numerous and varied. Some are accepted as inevitable, but as new products, materials and techniques become available the possibility of their elimination improves. The plant engineer and designer must learn to look for and recognize the possibilities that these new methods offer, however unlikely they might at first appear.

The more conventional methods include:

1. Simplicity in design and detail.
   Mechanisms that are easily understood by the average maintenance tradesman facilitate maintenance.

2. Designed unit replacement.
   Replacement of a complete faulty unit is often quicker than repairing *in situ*. Downtime is reduced.
3. Designed standardization.
   See page 88 for the advantages of standardized parts, assemblies and machine units.
4. Improved lubrication systems.
   See below.
5. Improved safeguards against dirt, grit, water, etc.
   Flexible protective covers over slideways. Wipers to clear away chips, grit, dirt, etc. from sliding surfaces. Flexible gaiters round sliding rods.
6. Hard facing to resist wear and abrasion. Applications:
   Valves and seats. Guillotine blades; punches; shears; cutting tools. Cams; tappets. Mixer blades. Shovel and dragline teeth; digger teeth. Crusher jaws; rolls.
7. Improved arrangements for inspection.
   Inspection doors and windows to view workings of mechanism without dismantling unit. Plug holes for insertion of endoscopes, inspection mirrors, fibre optics, etc.
8. Improved arrangements for test and fault location.
   Test points for module or unit testing. Pressure tappings. Built in temperature probes.
9. By-Pass and cross-over connections.
   To enable maintenance to be carried out 'on the run' and thereby reduce downtime.
10. Identification and marking of valves, pipes and wiring.
    To assist tracing and fault finding.
11. Better accessibility for maintenance work.
12. Better instruction and service manuals.
    Many plant failures and excessive wear and tear are the direct result of misuse. This is not intentional but often due to ignorance on the part of the operator. Training in the correct methods of usage could, in this case, be more accurately termed *training-out maintenance,* so could courses, often specialized, that help tradesmen do a better job.

3.2. Planned lubrication

The root cause of many machine failures is incorrect lubrication – incorrect grade of lubricant, incorrect type of lubricant, too much lubricant, too little lubricant or infrequent, or absence of, lubrication.

In these circumstances, correct lubrication assumes the role of *preventive maintenance,* (BS definition – *work which is directed to the prevention of failure of a facility*). Lubrication should be considered as important as *planned maintenance* and given the same forethought and control as other maintenance functions; it cannot simply be left to machine operators to organize. Many companies have realized the benefits ensuing from correct planning of the lubrication function by setting up a properly planned, organized system, employing personnel whose specific responsibility is the lubrication of the machines.

The approach to any system of planned lubrication is similar to that of planned maintenance described in previous chapters. The same general principles apply:

What is to be lubricated?
How is it to be lubricated?
When is it to be lubricated?
Is the lubrication effective?

All systems should ensure that:

1. Every machine or item specified is correctly lubricated.
2. The correct type, grade and quantity of lubricant is applied.
3. The correct lubricant is applied at regular specified periods.
4. The lubrication is recorded to indicate the effectiveness of the function.

Each system must be adapted to suit circumstances and needs, but it must be simple to operate, requiring the minimum of clerical and paper work. Many systems not only fulfil the basic objectives, but also present recorded information in a manner that can be used for historical and statistical purposes. Thus a number of secondary benefits are obtainable which increase the usefulness of the system:

1. Establishes responsibility for who carries the work out.
2. Establishes responsibility for the work actually carried out, or omitted.
3. Establishes the oiler's work load – optimizes available labour.
4. Establishes how the lubrication should be carried out – the method and route.
5. Reduces lubrication errors and omissions.
6. Highlights faults and unusual circumstances immediately.
7. Improves control of stocks and ordering procedures.
8. Establishes a basis for statistical analysis of lubricant consumption.
9. Rationalizes use of lubricants.

Planned lubrication and control methods are usually more easily applied, and the results and advantages become apparent much sooner, than other planned maintenance activities.

The initial step in any planned lubrication system is to carry out a comprehensive survey, listing all the items of plant to be lubricated and the complete lubrication requirement of each item. This information must be both complete and accurate, as it is upon this that the operation is based.

Details compiled during the survey should include:

1. Type of machine to be lubricated.
2. Identification number of the machine.
3. Location of the machine.
4. Detailed breakdown of the lubrication requirements, such as,
   section or position to be lubricated,
   lubrication point reference number,
   number of lubrication points,
   type and grade of lubricant to be used,
   method of application,
   frequency of application,
   frequency of lubricant change or repacking,
   capacity of lubricant reservoir or change,
   any other special considerations.

Much of this information will be available in the manufacturer's servicing manuals. This will reveal that a large number of different makes and grades of lubricants are currently being used. Most machine manufacturers quote alternative makes and grades of lubricant that are suitable for their machines; it is often possible to rationalize the range required. Alternatively, most oil companies are prepared to examine the problem and submit a short list of their own lubricants that would be suitable for the needs of the entire factory. The fewer lubricants required, the simpler will be the system, while possible errors in dispensing and application are reduced. Benefits are also obtained in bulk buying, storage and dispensing.

Other relevant observations made during the survey should be noted:

1. Inaccessible lubricator points.
2. Guards that must be removed to gain access to points.
3. Inaccessible drain taps.
4. Lubricator points dangerously positioned.
5. The different types of lubricator points fitted.

Ideas which make lubrication easier, quicker and reduce the risk of mistakes or omissions should be considered in the plan. A method study of the lubrication arrangements will highlight the advantages to be gained from:

Automatic lubrication systems.
Centralization or grouping of lubricator points.
Extension of lubricator pipes to facilitate access.
Standardization of lubricator points (either throughout the plant or for the different grades of lubricants).
Larger feed cups or reservoirs (require less frequent filling).
Marking lubricator points to indicate the frequency, type and grade of lubricant to be applied.

A standard code for lubrication symbols has been established by PERA• in conjunction with a number of leading oil companies. By the use of coloured shaped symbols it is possible to provide a simple visual indication at the point of application of the frequency, the type and the grade of lubricant required (Fig. 3.1). An individual shape denotes the frequency of application, the colour of the shape identifies the type of lubricant, while a number on the shape indicates the grade (or viscosity). These shapes are available in coloured plastic so they can be attached readily to the machine.

STANDARD CODING FOR LUBRICATION SYMBOLS

FREQUENCY CLASSIFICATION: SHAPE-FREQUENCY
Daily
Weekly
Monthly
Scheduled—for frequencies other than those above.

LUBRICANT CLASSIFICATION		LUBRICANT CLASSIFICATION GRADE NUMBERS: VISCOSITY RANGE — REDWOOD SECS										
COLOUR	OIL	35-40	41-45	46-55	56-75	76-100	101-130	131-170	171-300	301-450	451-600	601-1000
Red	Lubricating—Light	1	2	3	4							
Black	Lubricating—Medium					5	6	7				
Orange	Lubricating—Heavy								8	9	10	11
Blue	Hydraulic	1	2	3	4	5	6					
White	Slideway				4/5		6/7		8/9			
Red stripes on white ground	Special Purpose This applies to any oil outside the above classification.	Grade numbers do not apply. Sequence numbers are applied to differentiate between types of special purpose oils.										

COLOUR	GREASE	N.L.G.I. No. 1	2	3	4
Yellow	Lithium	1	2	3	4
Pink	Soda	1	2	3	4
Green	Lime	1	2	3	4
Yellow stripes on black ground	Special Purpose This applies to any grease outside the above classification	Grade numbers do not apply. Sequence numbers are applied to differentiate between types of special purpose greases.			

Supplies of symbols and further information on the code can be obtained from: PERA, Melton Mowbray, Leics.

Grade numbers are black or white dependent on background colour.

3.1 Standard coding for lubrication symbols

When the make, type and grade of lubricant, together with the method of application, have been finalized, the information that sets out the complete lubrication requirement of each machine or component should be recorded on an individual sheet or card (Fig. 3.2). These individual lubrication schedules should be compiled to form a lubrication register containing the details of all the machines to be included in the plan.

•Production Engineering Research Association, Melton Mowbray, Leics., England.

The information contained in the lubrication register, enables us to draw up a programme of periodic tasks. With careful planning and routing, a realistic workload can be established for each oiler. If 70 to 80 per cent of his time is spent on programmed work the remainder of the time allows for any unscheduled and contingency work that may arise.

In many factories, the responsibility for electric motor lubrication rests entirely with the electrical maintenance department. Such cases should be noted on the lubrication schedule or a separate schedule of these items should be maintained. A separate programme of work must also be issued to the appropriate department to ensure that the work is carried out at the predetermined time.

MACHINE: ____________
NUMBER: ____________

LOCATION: ____________

Mobil LUBRICATION PROGRAMME **machine register**

	ITEM	NO. OF POINTS	LUBRICANT	APPLICATION METHOD	INSPECTION/ APPLICATION FREQUENCY	CHANGE/ REPACKING FREQUENCY	CAPACITY IN GALS./LBS	
A								
B								
C								
D								
E								
F								
G								
H								
J								
K								
L								
M								

FORM 8022(6-70)

3.2 Machine register card

Having decided what machines are to be lubricated, how and when they are to be done, we now have to convey this information to the greaser. A system, adapted to suit local conditions and requirements, will be needed.

A simple system indicates only the machine, the date and the type of service (e.g. Service A, B or C; daily, weekly or monthly) to be carried out. At the other extreme there is the fully computerized system which is capable of automatically providing complete 'print-outs' of each task to be done as well as analysing and evaluating the results. Several different types of systems are described at the end of the chapter.

Most breakdowns stem from small beginnings. Frequently, these beginnings are of a most elementary and easily detectable nature – chaffed pipes, frayed belts, small oil leaks, to name but a few – but they are overlooked until they fail.

With a little training and instruction, the greaser is able to look for early signs of these defects, together with any other unusual conditions – hot bearings, reduced oil pressure, unusual oil consumption – that were observed on his daily rounds. There should be provision in the system for him to report such circumstances so they receive early corrective attention.

All tasks can be accomplished with greater efficiency if the right tools and equipment are used – lubrication being no exception. The provision of a suitable lubrication trolley containing the essential requirements saves a considerable amount of the greaser's time and enables him to carry out his work more efficiently. It transports his supplies and ensures that all the necessities are at hand, thus reducing the frequency of returns to the stores for replenishment. The trolley need not be elaborate – a simple platform truck would suffice, provided it was large enough to hold several oil containers (each with an easy means of dispensing), a range of measuring cans, oil cans, grease guns and a compartment for clean lint-free rags, or any other small spares that may be required.

To minimize mistakes, all containers, oil cans and grease guns should be marked with a colour code to indicate the type and grade of their contents – the respective colours being the same as the symbols attached to the machine lubrication points, details of which were discussed earlier.

Means should also be provided to drain sumps and reservoirs. A simple arrangement can be constructed from a platform truck upon which there are empty drums (to receive the old oil), lengths of hose and a hand suction pump. Where the capacities of the reservoirs are large an electric pump may be fitted. If required, specialist equipment, which may range from a simple tank and pump to the complete set up of filters, separators, tanks, pumps, etc., may be purchased.

If further information or assistance on a lubrication problem is required, consult the oil companies, who will offer their expert advice.

3.3. Labour relations

When the principles of planned maintenance have been accepted by the management and the decision taken to proceed with its introduction, the supervisors and staff of the maintenance and production departments should be fully consulted so their active co-operation and participation is guaranteed. Unless complete co-operation is obtained, the scheme will not function successfully.

Planned maintenance will create considerable changes. New working methods, new techniques, new ideas and new procedures may all be introduced. In many industrial situations, there is often a strong prejudice and resistance to change – new ideas are difficult to get accepted or are regarded with deep suspicion. Opposition to new innovations may range from complete indifference:

'We've been doing it this way for years – why change now?'
'Why change – we're doing alright?'
'It wouldn't work in our factory.'
'We've managed without it so far,'

to downright hostility:

Non co-operation.
Work to rule.
Industrial disputes.

The basic causes of these attitudes and reactions may be lack of enthusiasm, low morale, frustration in the former case or fear of reduction of wages, loss of security or unemployment in the latter instance.

Whenever changes are proposed which influence employment or working conditions, the primary concern of the individual is a personal one:

'Will it make any difference to *me*?'
Will it affect *my* future employment?' (Redundancy.)
Will it affect *my* wages?'
Will it affect *my* working conditions?'

Secondly, the broader issues will be questioned:

'Why is the company doing it?'
'What will the company get out of it?'
'What is the *real* reason?'
'Exactly how will the scheme operate?'
'What changes are proposed?'

Initially, the maintenance staff may resent the fact that the maintenance schedule instructs them exactly:

What to maintain.
How to maintain.
When to maintain.

They will complain that the traditional skill has been taken out of their work and that initiative is no longer required. There may be a similar resentment from the production staff.

In the absence of correct and proper information from management, rumours will abound, fears and suspicions will multiply and labour relations will become strained. In this climate, the smooth introduction and subsequent operation of any new scheme is bound to fail. Whenever changes are contemplated the facts must be given, the questions answered and agreement reached *before* the measures are introduced. Failure to observe these preliminaries could lead to problems that blow up into major issues, causing drastic modification or even complete abandonment of the original idea.

The spirit with which new ideas and changes are accepted will depend upon the relationship that exists between management and staff. A good relationship is based upon mutual confidence supported by an efficient means of communication. Methods of communicating information differ but whatever is adopted a correct *chain of communication* must be established and adhered to. This procedure ensures that the information originates and circulates in a predetermined manner, so that it is made available to the right person, or persons, at the right time and everyone is 'in the picture' in the correct order of precedence.

The presention of information is equally important. It is of little use displaying a notice on a board expecting everyone to read, understand and agree with it. Different people interpret the same facts in many ways and arrive at widely differing conclusions. Not everyone has the same point of view or the same scale of values. Consequently, it is vital that the company's intentions and proposals are made absolutely clear so that they are not open to misinterpretation. The responsibility for communicating this information should be delegated to a person who is familiar with the situation and can foresee its possible future effects.

His task is to:

(a) sell the idea,
(b) convince the staff that the new methods are unquestionably better, and
(c) educate the staff into new ways of thinking.

The most effective way to do this will depend upon many different factors, but to 'put over' the information successfully he must be prepared to spend considerable time and effort discussing, explaining and answering questions, all to the satisfaction of the people involved. This may require attendance at works committee meetings, organizing lectures, short courses, film shows, exhibitions or poster displays. Every channel of communication should be used to ensure that the information is made available to each individual person concerned and that it does not become distorted in its passage. Ultimately, all persons should not only fully understand the idea, but should also completely agree with and accept it. The following notes may be helpful when preparing a communication:

1. Personal contact is much better than the impersonal approach.
2. Use simple words and phrases that are understood easily by all levels of education and intelligence.
3. Be clear, precise and concise.
4. Be systematic, deal with the subject matter step by step in easy, progressive stages.
5. Ensure the information is correct and accurate.
6. Be extremely careful in expressing opinions. (They have a nasty habit of rebounding.)
7. Be completely honest and sincere about the situation, especially the parts which may be unpopular.

8. Use visual aids to support and confirm your efforts.
9. Do not expect to achieve overnight results.

For communications to be of any real value there must be a two-way flow of information, and a free exchange of ideas between all parties. This offers the opportunity to check that the impressions created are, in effect, the ones intended and it enables each side to appreciate the other point of view. Difficulties not apparent to the planners may be noticeable to persons who will be required to carry out the work; joint discussion at the proposal stage enables the necessary modifications to be made before they become major issues.

Grievances, misunderstanding, problems concerning personal attitudes, and organizational and operational snags are bound to arise with any new scheme, but when these differences are discussed openly in an atmosphere of mutual understanding in which a genuine desire to overcome them prevails, the solutions are more satisfactory, longer lasting and are much more acceptable to all concerned.

During the course of work, most persons think up ideas and improvements that will help with their jobs. Sometimes these ideas are brought to the notice of the management and are of sufficient merit to be applied successfully. Many other good ideas slip by unnoticed. Channels should be available to allow suggestions to be received and considered. If people have a sense of participation in the development of ideas, improvements and schemes, they will be more inclined to accept and adopt them.

3.4. Store keeping and stock control

Stock control is a clerical system for controlling the movement and availability of stock.

Store keeping is the physical control – receiving storage and issuing – of stock.

Adherence to the maintenance programme will, in many instances, depend upon the availability of spares. The absence of a certain spare can disrupt the whole programme or, in the case of a breakdown, prolong the downtime. To overcome this, large stocks of spares are often held – this policy can prove equally expensive. Storage space is usually at a premium, clerical and manual labour is needed to administer the paperwork, handle and check the stocks, and deterioration or obsolescence can occur during storage. All the time that spares are held in the stores they represent capital lying idle and incurring overhead costs. Thus a balance between the extremes of low stock and high stock levels must be established, one that will provide an acceptable measure of safety and yet is not too costly.

In most organizations, the spares and stores required by the maintenance department can be divided into the following categories:

1. Spares.
   (a) Programme spares.
   Spares and materials required for the scheduled or planned work contained in the maintenance programme.
   (b) Break-down spares.
   Spares held as an insurance against the unexpected breakdown of parts vital to the operation of the production plant. As planned maintenance progressed so the number of these spares would decrease.

Some of the spares in both the above categories might be reconditioned units ready for re-use.

2. General maintenance stock.
   Valves, piping, flanges, electric cable, conduit, switches, etc.
3. Consumable stores of a minor nature.
   Nuts, bolts, packing, jointing, etc.
4. Tools and equipment.
   Specialized tools and equipment, lifting gear, chains, etc.

Invariably, it is the items contained in categories 1(a) and (b) that have the greatest effect on the maintenance programme and also the ones that incur the greatest costs. By Pareto's Law (page 132), it is the elements in this category that should receive the most attention.

Stock control ensures that sufficient but not excessive spares and materials are available when they are needed. Before stock control can be instituted, we must decide what spares should be stocked and the stock level of each. Manufacturers usually issue lists of recommended spares with their equipment. These, in conjunction with information extracted from the maintenance schedules and the maintenance programme, offer a realistic basis from which to start. A point to remember: because, say, three identical units are installed, it does not necessarily mean that a set of spares must be stocked for each; two or even one set may well suffice.

Effective stock control relies on accurate, up-to-date, readily available information – it is the stock record card that provides the elements of this service. Each item or spare has its own record card which clearly indicates:

1. The quantity in stock.
2. The minimum stock level below which it is not safe to fall.
3. The maximum stock level above which it is not necessary or economical to rise.
4. Re-ordering point, the level at which new stock should be ordered. This takes into account the normal usage rate and delivery time so that stocks do not fall below the danger or minimum level.
5. Quantity to be re-ordered each time, usually to bring stocks up to the maximum stock level.

6. Quantities reserved for future planned work.
7. Outstanding orders to be delivered.

Figure 3.3 is an example of a stock record card accommodating all these points.

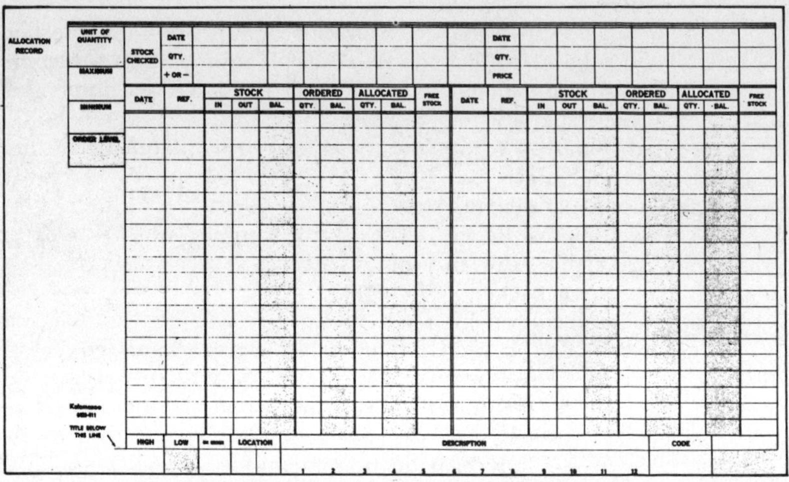

3.3 A stock record card (Kalamazoo loose leaf records)

It is pointless to operate a stock control system if it is not matched by an efficient storekeeping procedure. The control machinery becomes an expensive, clerical exercise and the stock records a valueless sham if spares, when needed, are found to have been damaged or 'lost' in storage. Store keeping entails:

1. Receiving and booking in the goods. Returning faulty goods and packages when necessary.
2. Storage in conditions that protect the stock from damage, deterioration and pilfering.
3. Keeping records of stock.
4. Requisitioning supplies to maintain stock levels.
5. Periodical checks for inventory purposes.
6. Issuing stock as and when necessary.

Whatever system of storekeeping and stock control is used, a physical check of incoming goods should always be made to ensure that they conform both in quantity and specification to those ordered, and that they have not been damaged in transit.

3.5 Standardization

Standardization is the removal or prevention of needless variations or irregularities within a group of related items. Its object is to establish a uniform pattern that conforms to an approved standard.

Standardization is fundamental to modern industry. Without it, industry could not exist in its present form; it affects every industrial activity. It is as relevant to operational and administrative procedures (procedural standardization*) as it is to material objects (physical standardization†).

Some forms of standardization are so fundamental that they are taken for granted – driving on the left, red for danger, pipe fittings, electric lamp connections, are but a few examples. Every company accepts them automatically, indeed, not to do so would prove extremely expensive and inconvenient, but not every company actively or conscientiously pursues standardization as a means of greater economy or efficiency. Much has been published already about the many different aspects of standardization, so this chapter will only be concerned with its application and effect upon the maintenance function. Many of the points raised are applicable equally to other sections of industry.

There are disadvantages as well as advantages in standardization, but generally the advantages outweigh the disadvantages. The usual criticism of standardization is that it allows only limited individuality or latitude and does not cater exactly for particular conditions in the same manner as a specially tailored product, machine or system. The substance of this is not disputed but the 'price' paid for this flexibility or the extra, often marginal, benefit requires very close scrutiny. The restrictions imposed by standardization are often its source of strength. Standardized methods attempt to suppress erratic or impulsive action and often personal judgement to effect uniformity. This may limit or even exclude alternatives but it can relieve management and personnel alike of many day-to-day, time consuming decisions.

The principles of standardization can be applied to most installations. A company operating a number of similar machines may achieve complete standardization by adopting a single design or model, but this is not usually possible if the various processes are highly individualistic, purpose-designed and built to produce different, often sophisticated, products. However, even in the latter case, the designer can standardize auxiliaries, units, electric motors and switchgear, valves, instruments, components, and so forth, that are common to all processes.

Some of the many advantages to be gained by standardization are listed below.

Advantages of physical standardization:

1. Selection of equipment is simplified.

*Procedural standardization: Standardizing the mode of action or technique – human or otherwise, e.g. Accounting procedures and systems, purchasing procedures, filing systems, safety procedures, process plant operation, and methods of test, inspection and maintenance.

†Physical standardization: Standardizing those properties that can be defined in positive, quantitive, measurable, substantive terms, e.g. Type, model, manufacturer; dimensions, capacity, speed, etc., and chemical composition.

2. The types, variety and numbers of spares held in the stores are reduced.
   (a) Less capital is tied up in stocks.
   (b) Less storage space is needed.
   (c) Stocktaking, storekeeping and purchasing are easier and less costly.

With respect to marine engineering it has been stated[4] that 'a class or unit of less than three requires as many base spares and as much organization as a class of twice as many.'

3. By standardizing on manufacturer's standard equipment, replacements and spares are less costly and their delivery quicker.
4. Facilitates increased interchangeability of modules between the different facilities.
   (a) Unit replacement, i.e. gear boxes, pumps, motors, tube nests, etc., becomes a practical and economic proposition.
   (b) The cost in both labour and downtime of overhauling or repairing a unit *in situ* can be very high. It is often much quicker and cheaper to fit a new or reconditioned replacement so that the old unit can be overhauled in the workshops and made available for re-use on another facility.
5. Fewer reference drawings, specifications and maintenance instructions are required.
6. A smaller variety of machines requires fewer specialized tools. But on the other hand, because of their greater utilization, it becomes more profitable to purchase special maintenance and test equipment instead of spreading the same cost over a larger number of general purpose tools to suit all machines.
7. Because fewer types of equipment are involved, familiarization is quicker.
   (a) Less time is required to train operating and maintenance personnel.
   (b) Operating and maintenance personnel get to know the characteristics of the equipment much quicker. The pattern of maintenance becomes more intimately known, faults are recognized, diagnosed and located much quicker.

Advantages of procedural standardization:

1. Operational uniformity is created.
2. Persons operate with identical methods and achieve consistent and comparable results.
3. Continuity of the system is not broken if persons are transferred.
4. Established procedures become routine and can be applied repeatedly without further attention.

5. The results from procedures and methods of inspection and test are always comparable.
6. Standardizing the quality of work or level of performance within acceptable limits ensures that:
   (a) the level is not unreasonably high and difficult to attain, causing unnecessary costs and wasted effort, or
   (b) the level is not too low so as to cause inefficiencies, premature or excessive breakdown, high scrap or reject rates.

The majority of companies have standardized procedures for most of their internal activities – accounting, purchasing, communication, budgetting, cost control, etc. All these procedures are accepted and enforced with uniformity, irrespective of type or size of branch or department. This, however, is not entirely possible with planned maintenance systems as there is no single system that is suitable for all situations. Each must be tailored to meet the individual circumstances. Consequently, the organization of a planned maintenance system at each branch is left to the discretion of the individual works engineer with the result that there are as many different maintenance systems as there are works engineers. When these engineers are transferred, perhaps promoted, between the various branches, they tend to supplant or modify any previously existing system, no matter how good, with their own favoured ideas. Such independent and inconsistent local arrangements are not conducive to continuity. Not only do they have a most unsettling effect upon staff, especially if transfers occur frequently, but they offer no planned common basis for comparison or progressive improvement.

If the company is to standardize on a single planned maintenance system that is equally applicable to *all* its branches then the approach must be one of *standardization with individuality* – an apparent contradiction of terms. All planned maintenance systems must satisfy certain basic conditions (see page 13), by standardizing the broad principles on how these conditions must be achieved the first term of the expression is satisfied, and by allowing the application of these principles to be arranged to suit the individual circumstances the second term of the expression is fulfilled. In practice, a company might well institute a standard basic system using the same documentation, filing, recording, and general procedure at all its branches but allow each works engineer sufficient, but not unlimited, scope to adapt it to suit the particular needs of the branch.

The initial choice of a basic system will be influenced considerably by complementary procedures and systems – job instruction cards, work progress and control systems, job records – that are already in use within the company. Where these systems and equipment have been installed by a specialist supplier – Kalamazoo, Remmington Rand, Roneo, Seldex, etc. – they are capable of being extended to include planned maintenance.

Such an extension of an existing system helps to consolidate the standardization process.

3.6. Pilot schemes

A pilot scheme is a preliminary, experimental trial of a project on a small scale.

This is probably the company's first venture into planned maintenance so the progress of the scheme will be under close scrutiny from all sides. Management will be looking for early signs of positive achievement to prove the scheme's value and to justify further support. The sceptics and the pessimists will be awaiting the first opportunity to shoot the whole thing down, while others will be sitting on the fence waiting to be convinced one way or the other. The organizers will be feeling their way, often harassed, but doing their best to smooth out snags and problems as they arise, then modifying in the light of experience gained. If the scheme is to survive, it must quickly prove its potential and thus satisfy management and convert the critics. Results in the early stages can 'make or break' a project.

The detailed organization and implementation of planned maintenance for a whole factory could take many months of hard work. Formulating the plan, acquiring the documentation, compiling the inventory, the facility register, the maintenance schedules, the job specifications for all the items is a time-consuming, monumental job with no measurable benefits. Even when the maintenance plan has been put into operation a further period of time will elapse before the results of the work become apparent. At first, not only must the planned work be done but additional work or replacement parts may be required to bring the equipment up to its specified operational standard. This extra, corrective work over and above the normal planned maintenance load could tax severely manpower resources, jeopardizing the whole operation. In all, considerable time, money and effort in excess to that previously incurred will be expended before the investment begins to show any return, during which time the scheme is a 'sitting duck' for the critics.

No matter how well the scheme appears to be planned, practical experience will show up its weaknesses, and once it is in operation improvements are not always easy to incorporate. Amendments or changes in documentation can be costly, and procedural and administrative changes may be difficult to implement, tending to cause misunderstanding and confusion as well as creating a false impression of insufficient thought and muddled planning. The introduction of planned maintenance simultaneously throughout the whole factory is courting trouble. It would be more prudent to start with a pilot scheme, using it to develop and prove the system before its adoption on a much wider basis.

Having evolved a satisfactory system it can then be extended gradually to include other sections of the factory.

Ideally, a pilot scheme should take place in an environment that is fully representative of the operational situation. A small self-contained section of the factory, a product or a process line, a small group of machines, equipment having a number one priority with a past history of failure or not attaining its required level of performance, would each provide a suitable starting point. However, if the results and the experience so gained are to be valid for large scale operational use, then the balance between the various elements – mechanical, electrical, instrumentation, lubrication – and their respective maintenance tasks should be typical for the factory as a whole.

By virtue of the small scale of a pilot scheme, the organizers can enjoy certain advantages that are not readily available to administrators of large scale operations. Fewer persons are involved so there is a greater opportunity for personal contact between the organizers and the shop floor operators. Such contact bridges the gap between 'them up there and us down here', it helps to establish a common bond with a common objective. It promotes an atmosphere of team spirit. Also, knowing that attention is being focussed on the scheme and that the outcome is dependent partly upon his efforts, the individual has a deeper sense of personal involvement. He will work for the success of the scheme because *he* is part of it. Elton Mayo in his famous Hawthorne experiments demonstrated the power of this psychological factor (see also Labour Relations, section 3.3).

A pilot scheme, being experimental, may receive certain privileges and allowed certain liberties which are advantageous. This is as may be, but its more usual and tangible advantages are numerous:

1. It is easier and quicker to organize and administer.
2. It is more manageable, easier to direct and control.
3. Results are evident much sooner.
4. It is easier and less costly to initiate modifications and amendments.
5. It is easier to visualize and observe the entire project, its operation and its progress, and to keep in close touch with every aspect of it.
6. Expensive documentation can be dispensed with, office duplicated paperwork will suffice until the system is perfected and goes into full operational use.
7. Greater freedom for experimentation is possible and is accepted as a part of the evolution process. Reputations are not necessarily at stake if the results of the experiments do not come up to expectation.

Although it might be very desirable to produce a fully representative tried-out model scheme before implementing planned maintenance on an operational basis, there are occasions that necessitate a slightly different approach. To many companies, the idea of instituting planned maintenance is conceived, not out of any carefully calculated assessment of the potential benefits, but from a faint and belated hope that it might correct a deteriorating situation. It is only when frequent breakdowns create acute production problems – high scrap rates, poor quality, bottlenecks, late completion dates, lost orders, a maintenance department desperately endeavouring to cope with breakdowns as they occur – that the thought of planned maintenance ever arise. Even then it is considered indifferently and with an attitude of resignation – 'we've tried everything else, might as well give it a try, what is there to lose?'

In a situation of this type, speed is vital – quick results are needed, the maximum impact in the minimum time. Time may not allow the perfecting and proving of a pilot scheme before operational use; out of necessity introduction must be on the job, applying the elementary principles and learning from experience and mistakes. Initially, attention must be given to the key items – those that exert the greatest influence and cause the greatest disruption to the production process. No matter how crude the schedules are initially, they are a start and can be refined later, in the light of experience. Mistakes will be made and are to be expected but nothing succeeds like success. As the situation is controlled and the results emerge, so the system can be extended to include other items with a lesser priority rating.

For reasons discussed previously (Labour Relations, section 3.3), there may be initial opposition to planned maintenance, but usually there is no such opposition to planned lubrication. A planned lubrication scheme could well pave the way for planned maintenance proper.

In many respects, there is a close similarity between the application of planned maintenance and certain sections of the Factories Acts (see extract of Factories Acts, Appendix 2). Both require particular items of equipment to be examined and maintained at set periodic intervals, and records to be kept for future reference. In addition, setting up a system to carry out the requirements of the Acts is somewhat easier – the equipment involved, the type of maintenance and examination, the frequency at which it must be done, are all specified by law. Management is under a legal obligation to comply, tradesmen and production staff acknowledge that it is done in the interests of safety – their safety. Consequently, there can be no valid reason for opposition to the introduction of a system that endeavours to ensure complete compliance with statutory regulations. In some instances, such a system becomes an urgent necessity (see Factories Acts Project, page 208). Having established and gained experience of a system to accommodate the requirements of the Factories Acts it could then be extended to include other types of equipment, as and when the climate is more receptive.

3.7. Procedure for implementing a planned maintenance scheme

A summary of procedure is tabulated below and illustrated in Fig. 3.4.

Phase 1. Statement of management's policy:

1. Decision by management to either,
   (a) investigate the application of a planned maintenance scheme, or
   (b) organize and operate a planned maintenance scheme, together with the scope of the scheme:
      the whole factory,
      section or department of the factory,
      a particular production line only,
      one type of machine only,
      if works buildings, services, fire appliances, lifting gear, transport, etc., should be included,
      if incentive bonus scheme to be considered for maintenance personnel.
2. Appoint person to be responsible to either investigate or organize and operate scheme – as decided in item 1.
3. The person appointed given a comprehensive brief detailing his duties, responsibilities and authority:
   (a) Management's decision taken in item 1.
   (b) Extent of responsibility and authority.
   (c) Objectives of scheme – management's requirements, such as:
      higher production levels
      lower maintenance costs
      higher plant utilization
      reduction of breakdowns
      better utilization of maintenance staff
   (d) Target date for either report to be ready or scheme to commence functioning
   (e) Cost limits
   (f) Staffing, clerical staff, maintenance staff.
4. Department managers, foreman, supervisors advised of the management's intention regarding the proposals of a planned maintenance scheme. These persons to be kept informed as the scheme progresses through the various stages in order to obtain their full co-operation.
5. Maintenance and production staff advised of the management decision regarding the proposals and objectives of the planned maintenance scheme. The normal channels of communication used, works committees, etc., if necessary, the person responsible for the scheme attending to answer any relevant questions.

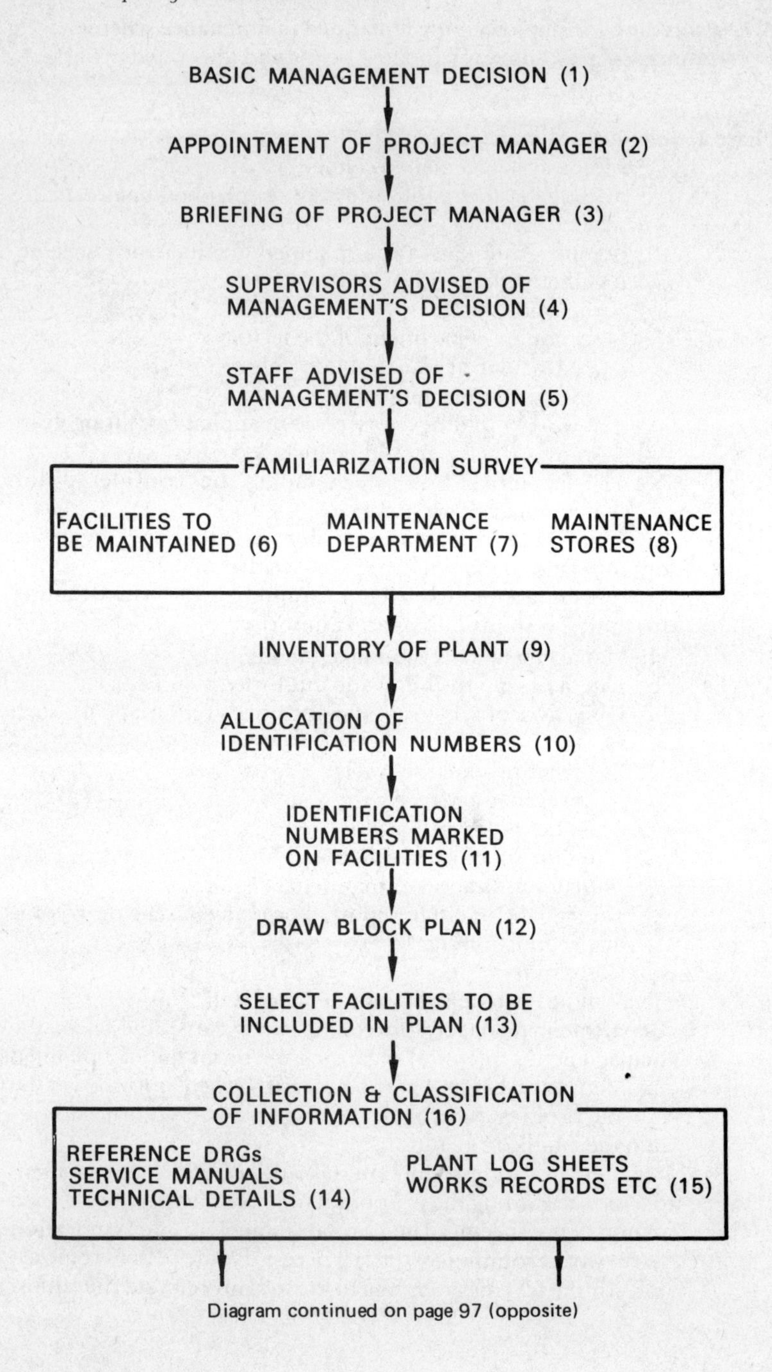

Diagram continued on page 97 (opposite)

Continued from page 96 (opposite)

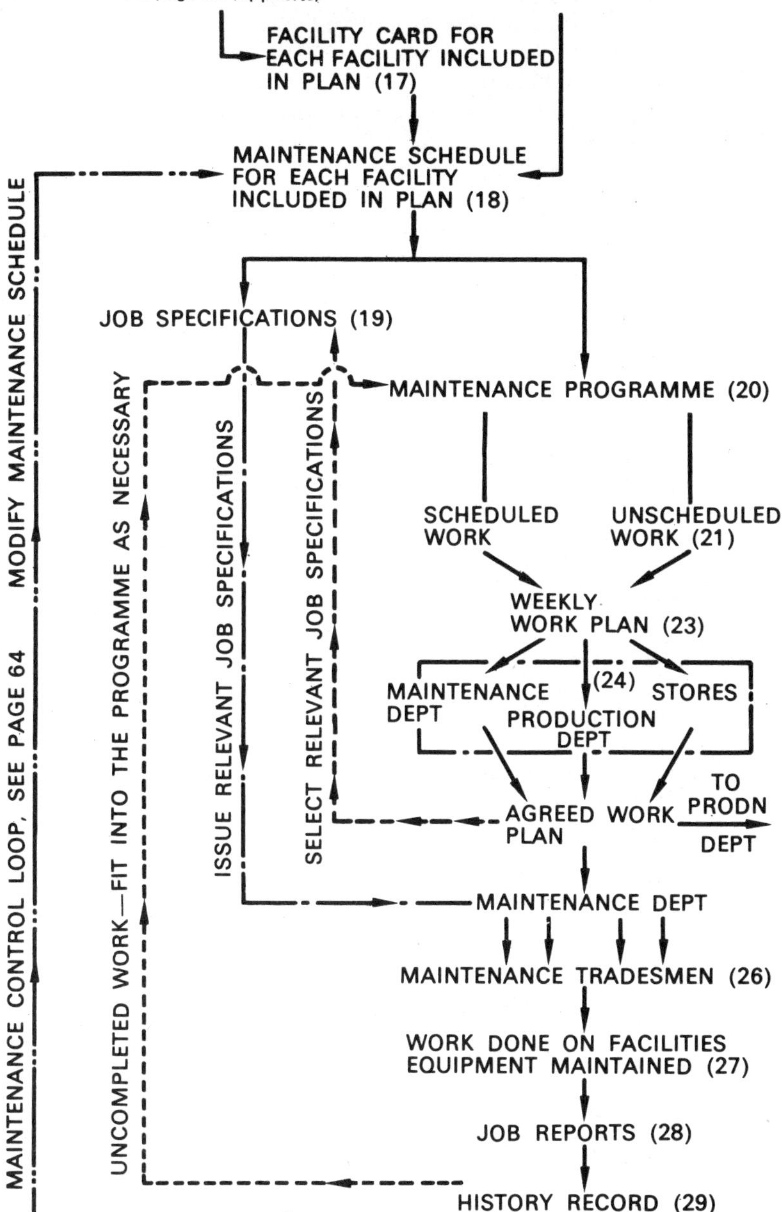

3.4 Outline procedure for a planned maintenance scheme (numbers in brackets refer to the item numbers in the summary of procedure)

Phase 2. Preparation

6. Make initial familiarization survey of factory, plant or section concerned.
   Note and summarize:
   methods and system of production,
   types of machines, production hours,
   shift work system, manpower,
   time available for maintenance.
7. Make initial familiarization survey of maintenance department
   Note and summarize
   resources and limitations:
   manpower – type and number of tradesmen,
   machines and equipment available,
   hours of working.
8. Make initial familiarization survey of maintenance stores.
   Note and summarize:
   method of storekeeping,
   type of stores held, ordering procedure,
   stock control methods, manpower,
   mechanical handling methods.
9. Carry out a physical inventory of the plant contained within the area or section of the factory concerned.
10. To each item entered in the inventory allocate an individual identification number.
11. Mark on each item of plant its identification number by a standard method which is clearly visible.
12. Draw a block plan of the area under consideration showing location of all machines and items of plant.
    Mark up the drawing with the identification numbers against the respective items.
13. Select facilities to be included in the plan.
14. Obtain reference drawings, lay-outs, technical details and service manuals for all items of plant to be maintained.
15. Study past log sheets of plant operation, works orders, stores requisitions and orders. Call upon unrecorded experience.
    Note and summarize:
    frequency and types of breakdowns,
    frequency and types of replacements,
    major components used for repairs,
    length of time and cost to carry out the various repairs, inspections and breakdowns,
    falling off in efficiency over various periods of running time,
    frequency of statutory inspections and tests (i.e. boilers, pressure vessels, lifting gear),
    allowable limits of wear, etc.,

This collection and classification of information facilitates a closer and more intense study of the various maintenance problems. It can also indicate inefficiencies and lead to new lines of thought.

16. After a careful study of the information obtained from items number 14 and 15 decide the details and information to be recorded, the exact description of the work to be carried out, the frequency, etc.

Phase 3. Preparation of documents

For each item of plant to be maintained there should be:

(a) a facility record card
(b) a maintenance schedule record card
(c) a series of job specification cards
(d) a history record card.

17. An individual facility record card is made out for each item of plant to be maintained. Full technical and constructional details are entered, and the card filed in the appropriate section of the facility register.
18. For each item of plant entered in the facility register, a maintenance schedule record card is completed. This card details the maintenance operations and their frequency.
19. From each maintenance schedule record card extract the various individual maintenance operations and detail each on a separate job specification card. Each operation having an individual job number which is indicated on the job specification card.
20. Set out a programme of operations showing when each operation is to be carried out. Plan the operations so the work load is spread evenly over the whole year.
21. Make arrangements in the system to accommodate unscheduled work which may arise.
22. While the system of documentation is being set up, the following work should be proceeding:
    (a) Review and overhaul store keeping procedure, institute stock control if necessary. Bring stores up to the required levels.
    (b) Review maintenance labour force – type of trades, etc. Note any shortages or overstaffing in particular areas. Carry out any training necessary.
    (c) Review lubrication methods.
    Bulk storage – centralization – automation.
    If required, oil companies will advise.
    (d) Review and check the equipment, tools and machines required to carry out the maintenance work.

(e) If an incentive scheme is to be operated for maintenance personnel, the principles and details to be established together with their acceptance.

(f) If specialist contractors are to be employed for particular items of repair or maintenance, make the necessary pre-arrangements, (e.g. chemical cleaning and de-scaling; instrument engineers; electricians; non-destructive testing). Ascertain:

The amount of notice required; their availability.
The equipment to be supplied by the customer.
The power to be supplied by the customer.

Ensure the contractor is informed of the extent and details of the work together with the time available to carry it out. Obtain the contractors written confirmation on all necessary points before commencing the scheme.

(g) Investigate the implementation of standardization, work study, new methods, materials or equipment to facilitate greater maintenance productivity.

Phase 4. Operation of the scheme

However simple or elaborate the maintenance scheme is, an effective control system (documentation, etc.) must adhere to a pre-determined route. All persons concerned must be familiar with the system and follow it.

23. Approximately 10 days before commencement of the maintenance week concerned:
    (a) From the maintenance programme extract and list all the maintenance work to be carried out in that particular maintenance week.
    (b) List also the unscheduled work to be carried out in that particular maintenance week.
    (c) Combine the above two lists to form the weekly plan sheet.
24. Submit the weekly plan sheet to both the production and maintenance departments obtaining their mutual agreement and confirmation on the proposed maintenance work. Check with stores that spares are available. This ensures that the production machines will be available for maintenance at the required time. Also that any modification to the programme made necessary by revised production schedules can be accommodated.
25. Upon mutual agreement of the weekly plan sheet, select the relevant job specifications and approximately one week before the maintenance week concerned issue them to the maintenance department. This arrangement allows the weekly

work load of the maintenance department to be planned and allocated in advance.

26. The job specifications are issued to the respective tradesmen to carry out the work detailed.
27. Work carried out.
28. Feed back of information:
    The tradesmen upon completing the work detailed, or as much of it as time or circumstances will allow, completes the job report. This summarizes work done; work scheduled but not carried out; replacements used; other defects observed; results of tests, etc.
29. Information extracted from the job reports, supplemented by information from other sources, (operating log sheets, works requisitions, etc.) is entered on the history record card of the item concerned. If a history record graph is also used, the relevant data should be plotted.
30. When the information has been entered on the history record card it should be sighted by a person who is fully familiar with the technical operation of the plant item and can interpret the results of the maintenance. A study of the entries over a period of time will indicate any change in trend or behaviour. Certain aspects may be highlighted.
31. In the light of information noted in (30) modify the maintenance programme accordingly. Any scheduled work left not done – consider when it should be fitted into the programme, and enter it.

4
Supporting Services

4.1. Work study

Work study is the detailed examination and analysis of human effort in order to determine the best way to achieve a defined objective.

Its application need not be restricted only to industrial or commercial spheres but can be employed usefully in any situation which involves human activity.

To many, the term *work study* conjures up a picture of practitioners with stop-watches, notebooks and other paraphenalia, timing and recording the movement of each operator for the express purpose of fixing piecework or bonus rates. This rather slanted view is correct only up to a point, but is unfortunately the one which receives the most publicity. Work study is an umbrella term which covers many different techniques, some of which do involve timed observations but not always for the reasons quoted above. These techniques can be divided into two separate but interdependent groups:

Method study –

> The systematic recording and critical examination of existing and proposed ways of doing work, as a means of developing and applying easier and more effective methods and reducing costs (BS 3138 : 1969). '*How* the job should be done'.

Work measurement –

> The application of techniques designed to establish the time for a qualified worker to carry out a specified job at a defined level of performance (BS 3138 : 1969). '*How long* the job should take'.

The purpose of work study is to determine a more efficient way of utilizing money, manpower, materials and machines, the standards being defined in terms of work content or time taken, but usually cost. To establish the work content or standard time for a specific job, work measurement must follow method study in order to:

(a) ensure the job is being carried out in the best way, and
(b) define the method by which the job should be done.

Where several alternative methods are possible or suitable, the application of work measurement may be the only means possible of deciding which is the best one.

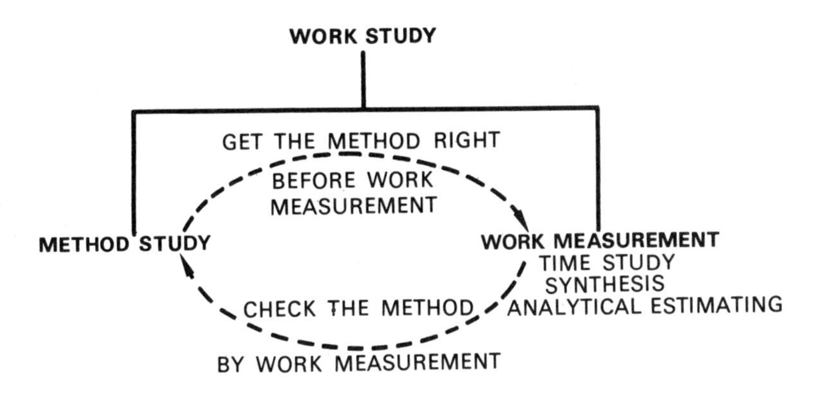

Work study is a specialist function which when applied professionally – incentive schemes, work payment, production and cost control – requires the services of a qualified practitioner. This expertise is essential for all forms of work measurement, but the other side of work study, namely method study, is within the scope of the layman, provided he recognizes his own limitations. As a parallel – builders, decorators, carpenters, automobile engineers are all specialists, but this does not prevent 'do-it-yourself' enthusiasts achieving passable results in their own homes and in the maintenance of their own cars; but even in these areas the wise recognize their limitations and leave certain jobs to the experts.

Likewise, it is not the purpose of this chapter to produce experts, but to explain the elements of method study so that the layman can employ the technique to bring about improvements within his own sphere of influence.

Method study

Method study is not simply finding a better way it is determining the *best* way. Conventional studies usually follow a well established pattern of:

Select – the job, work or problem to be studied.
Define – the objective.
Record – all the relevant facts.
Examine – these facts in a critical, systematic manner.
Develop – the *best* method.
Install – the method as a standard practice.
Maintain – the new method and review the results by regular routine checks.

No step should be omitted and each must follow in the sequential order listed above.

Figure 4.1 illustrates the basic procedural framework of conventional study, its virtue is in its ability to accommodate most situations, whatever their size and however diverse their nature.

Select – the work to be studied.

The subject of the study should offer a good return on the time, money and trouble spent on it. Success is assessed usually in terms of financial saving but studies are often undertaken to improve safety or working conditions. The following are typical situations that are likely to yield good returns as a result of method study:

Work with a high content of manual effort.
Work with a high material cost.
Work with a large continuous demand.
Recurring maintenance jobs.
Breakdowns or maintenance work that causes undue production stoppage.
Excessive downtime of plant and equipment.
Machines or processes producing a high scrap rate.
Low plant performance.
Bottlenecks.
Inefficient layout of plant and equipment – causing excessive movement of men and material.
Simplification of complex or difficult work into simple elements. (Difficult or complex work requires highly skilled operatives. There is also the greater possibility of overlooking and omitting some of the operations. Simplification results in the tasks being done easier, quicker, cheaper and more reliably.)
Unpleasant work – jobs of this type tend to be skimped or not done at all.
Fatiguing work – physically, mentally, visually. A fatigued person cannot work efficiently.
Dangerous or hazardous work.
Safety procedures.
Protective clothing and equipment.

Many studies are undertaken for reasons that are peculiar to a particular situation, process or machine.

Define – the objective.

Having selected the work to be studied the prime objective must be clearly defined and where possible specified in positive terms of what is to be achieved or what is the target. Typical objectives are:

The reduction of costs, time, materials, waste, scrap, downtime, human effort.
The improvement of maintenance services and methods, working conditions, safety practices, plant performance, plant layout, quality.

During the process of the investigation, a number of secondary benefits

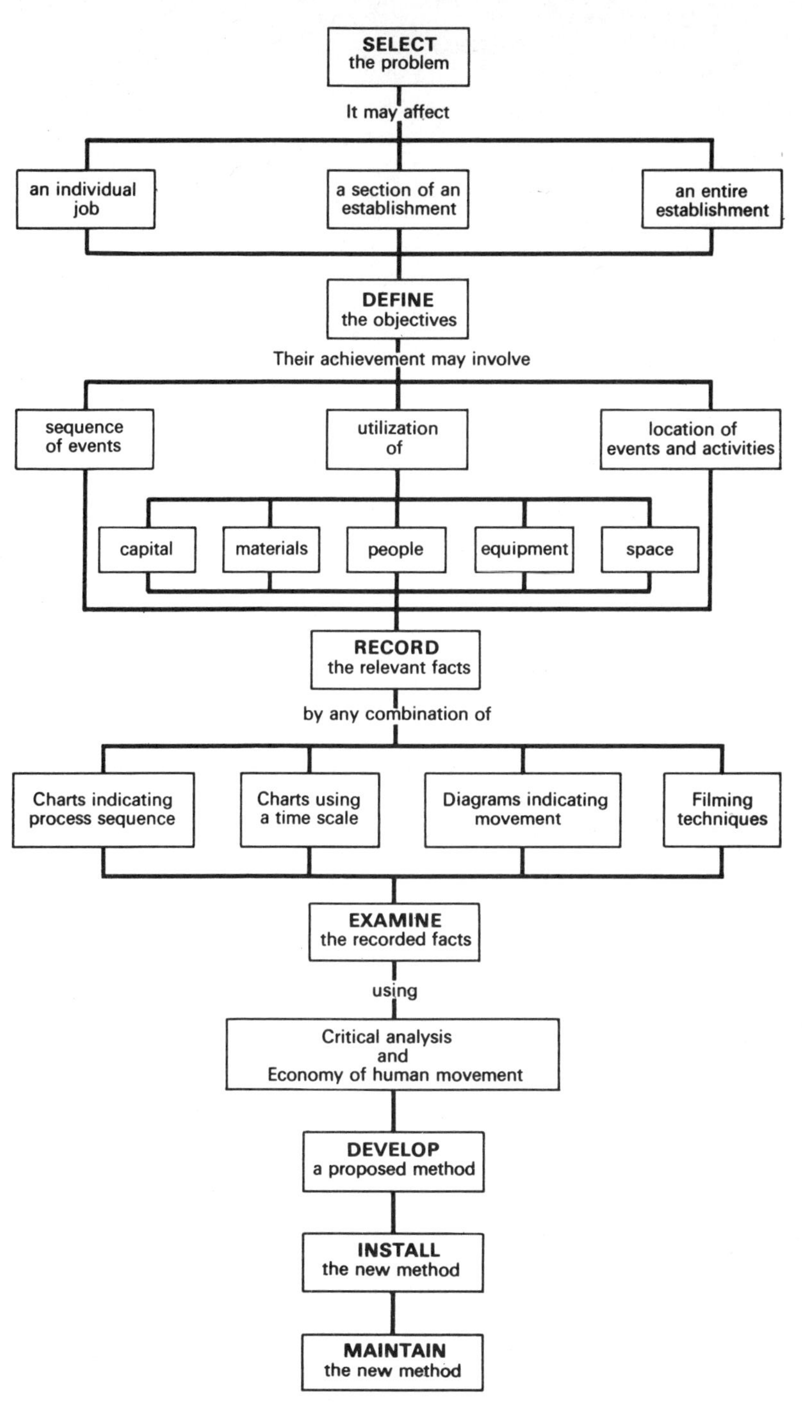

4.1 Outline procedure for a method study investigation

might become apparent, but these should not be allowed to obscure or displace the prime purpose – 'Don't loose sight of the *real* objective'.
Record – the relevant facts.

It is impossible to examine any activity without information. The more facts that are known about the subject being studied then the better will be the basis for future action.

The type of information recorded will depend upon the purpose and the nature of the study, but to be of any use it must be factual, accurate, up-to-date and presented in a readily accessible convenient form. The type of information usually recorded is:

The method, sequence of operations and how long each operation takes.
Number of persons and trades involved.
Tools, equipment and materials used.
Costs.
The snags and difficulties experienced in the operation.
Climatic or other environmental conditions that influence the situation.
The location, space and area used.
Performance and efficiencies.

This information might originate from a number of different sources:

Watching the work being done and listing the details.
Speaking to the persons doing the work.
Photographs, models.
Statistics – cost sheets, accounts, etc.
Charts, diagrams, graphs.

Examine – the facts:

The relevant information must be examined in sufficient depth to expose the true underlying reasons for every event and action. The whole exercise must be approached with a sceptical and inquisitive attitude, nothing must be taken for granted or assumed.

A widely used examination procedure is the question and answer technique. It consists of a series of basic questions designed to analyse methodically the constituent elements of a situation: (a) purpose, (b) place, (c) sequence, (d) person, (e) method – or means.

Each of these elements is then, in turn, considered from four different aspects:

1. What are the present facts – what is done now?
2. Each fact is challenged: Why is it done? Is it necessary? What does it cost, is it worth it? Does it contribute to a reduction in downtime, etc.?
3. What are the alternatives – what could be done?
4. Select the alternative for development – what should be done?

The whole procedure is in fact two processes – critical analysis and creative thinking. The latter stems from ideas and thoughts that are voiced or become apparent during the analysis stage. Every alternative should be followed up no matter how unlikely or improbable it might first appear, it could contain the germ of an idea which opens up new lines of thought. The whole procedure is summarized in Fig. 4.2, while Fig. 4.3 is a guide to its use.

Develop – the *best* method.

The most effective way of improving a task is to reduce or, better still, entirely eliminate the need for it. Failing this, the next best way is to simplify it. The selected alternatives should now be used to develop the best practical solution:

Review the ideas.
Combine, re-arrange and simplify the details.
Enlist the help of those doing the job and work out ideas with their co-operation.
Work out a phased development programme – short term, medium term, long term.

Install – the method as standard practice.

The point has now been reached to put the new method into practice. The proposals must be both acceptable to, and workable by, all parties concerned. The same principles of introduction apply as were discussed for the introduction of planned maintenance (Labour Relations, section 3.3).

Assuming agreement has been reached, the necessary arrangements for

	Present Facts		Alternatives	Action
Purpose A	What is done now	Why is it done	What else could be done	What should be done
Method B	How is it done now	Why is it done that way	How else could it be done	How should it be done
Sequence C	When is it done now	Why is it done then	When else could it be done	When should it be done
Place D	Where is it done now	Why is it done there	Where else could it be done	Where should it be done
Person E	Who does it now	Why that person	Who else could do it	Who should do it
			Possible ↑ alternatives	Selected ↑ alternatives

4.2 Critical examination sheet for method study

DESCRIPTION OF ELEMENT

A selected "Do" Operation

Reference....................
Page.........................
Date.........................

The Present Facts		Alternatives	Selected for Alternative Development
Purpose – WHAT is achieved? *Consider this operation in isolation (bear in mind the subject of the chart).* NOTE *What is ACHIEVED, not what or how it is DONE.*	IS IT NECESSARY? YES NO If YES – Why? *Reason given may not be valid.* True *reason must be uncovered.*	What ELSE could be done? *Can the achievement be ELIMINATED?* *Can the achievement be MODIFIED?* *All alternatives to the purpose should be stated including those which may require long-term investigation. The answer to this section is never "nothing"; there is always an alternative even if only the* non-*achievement.*	What? *Helpful to divide into short-term and long-term. Under long-term can go suggestions for future research and development.*
Place – WHERE is it done? *The location with reference to* (a) *Geographical position* (b) *Position within the factory, plant or area* (c) *Detailed position under* (b) *When appropriate, give reference to location and distance from preceding and succeeding activities.*	WHY THERE? *The reason for siting the operation there.*	Where ELSE could it be done? *Consider alternatives under each heading. Can working areas be combined or distances reduced?*	Where? *Where appears to be most suitable situation with present knowledge? Answer may be in relation to some other operation. Consider limitations of building design and services (steam, air) etc.*
Sequence – WHEN is it done? *What are the previous and subsequent significant activities and what are the time factors involved?*	WHY THEN? *The reason for the present sequence and time factor in the present process.*	When ELSE could it be done? *Can it be done either earlier or later in the process?* *If the sequence is fixed, can it be moved back to the previous operation? For example* "Immediately *after.*"	When? *As soon as possible in the process or immediately after the previous activity.*
Person – WHO does it? (a) *Grade, e.g. unskilled male* (b) *Employment, e.g. day worker* (c) *Name/s*	WHY THAT PERSON? *Reasons for choice under each heading.*	Who ELSE could do it? *All alternatives under each heading. Can a disabled person be employed?*	Who? *It may not be possible to select the individual without Work Measurement.*
Means – HOW is it done? *All relevant details are required of Material, Equipment and Operator engaged in the operation. Information should be tabulated as simply as possible under the following main headings.* (a) *Materials Employed.* (b) *Equipment Employed.* (c) *Operator's Method.*	WHY THAT WAY? *The reason should be investigated for* each *of the tabulated items under* each *main heading.*	How ELSE could it be done? *Investigate all alternatives for each main heading.*	How? *Decide the alternative for* each *item separately and knit together at development stage.* *Consider* safety. *Consider posture and environment of operator.*

tools, equipment, material and training must be completed so everything is ready when it is required.

Maintain – the new method:

Once the new method has been installed and is working on an operational basis regular checks are necessary:

(a) to ensure that the method is in fact used as it was originally planned and that the old ways or malpractices have not crept in,
(b) to smooth out any problems that arise through its use, and
(c) to ensure that the intended objectives are being achieved.

It may be judicious to modify the method in the light of operational experience.

4.2. Maintenance aids

Much of the work of preventive maintenance consists primarily of examination to ensure that certain components and assemblies are fit for further operational service. Even if it is found that their condition is satisfactory and there is no need for corrective action, considerable time is often spent shutting down, opening up for inspection and 'boxing-up' again afterwards. Inspections can be time-consuming and expensive in terms of both lost production and labour costs. Costs can often be reduced by the use of instruments or other diagnostic aids which make it possible to examine the parts or machinery without dismantling and, in some cases, while the plant is actually in operation. Such devices can be used also to monitor plant operation and give early warning of deterioration or faults before they lead to extended damage.

Some of these aids are described briefly, together with other items that could assist the maintenance engineer. Their cost ranges from one or two pounds in the case of the dye penetrant and temperature sensitive tapes to several hundred pounds for the more sophisticated equipment, but the cost is usually more than justified in the savings and protection that they afford.

Temperature indicating stickers

Flexible self adhesive strips which, when the temperature of the surface to which they are attached reaches a specified level an indicating panel on the strip turns black. Strips can be obtained to indicate one particular temperature or different temperatures within a set range.

Uses: To provide proof of excessive temperature exposure.
To detect temperature anomalies.
To measure temperature gradients.

Such strips can be affixed easily to potential thermal trouble spots – a rise in temperature is then clearly visible to passing persons.

Electronic thermometers (portable)

The instrument consists of a small, pencil-shaped probe carrying a heat sensitive semiconductor element connected by flexible cable to a pocket size indicator unit. Its accuracy and fast response to temperature fluctuations extends its field of application beyond that of conventional temperature measuring devices. It is capable of measuring solid, liquid, gas or surface temperatures.

Uses: Bearing temperatures; 'hot spot' location.
Checking air conditioning, refrigeration and heating systems.
Pipe and vessel temperatures.
Tracing temperature gradients and patterns.

Temperature probes

Temperature probes, thermocouples and resistance thermometers in a variety of shapes, sizes and mountings, built into the plant at inaccessible, thermally-strategic points, can give warnings of temperature anomalies before they cause trouble. By terminating the probe leads in socket outlet connections, a single portable indicator is sufficient to obtain a reading from any of the probes.

Uses: Temperatures of bearings; cylinder heads.
Winding temperatures of motors, generators, transformers.

Pistol thermometers

The pistol thermometer (Fig. 4.4) is a self-contained, battery-powered, portable instrument which measures the precise surface temperature of any object, moving or stationary, without surface contact. Temperatures of any liquid or solid – plastic, chemical, metal, rubber, ceramic, fabric, glass – can be measured.

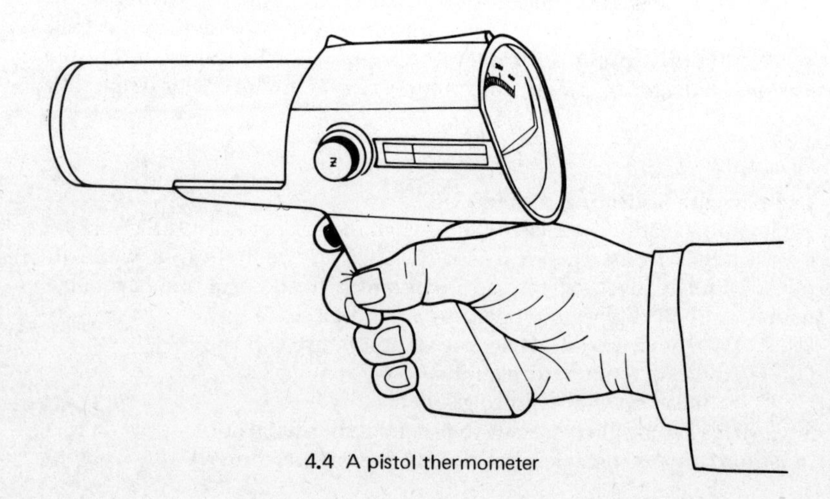

4.4 A pistol thermometer

The pistol, when pointed at an object, gives a direct temperature reading, temperature measurement being independent of distance. The operating distance can be as close as 152 mm (6 in) up to infinity, provided that the target fills the field of view. For specialized applications, instruments are obtainable with an operating distance of 12 mm to 100 mm (½ in to 4 in) and are capable of measuring miniature hot spots down to 2·5 mm dia (0·1 in dia.) in electronic circuit boards and similar components.

Various models are available, each is designed to measure temperatures within a specific range between 20° to 1650 °C (60° to 3000 °F).

The instruments facilitate spot checks for planned maintenance inspection and fault location. Dangerous, moving, electrically charged, untouchable or unreachable objects can be measured where other methods are uneconomical, impractical or hazardous.

Uses: Scanning surfaces for 'hot spots' (breakdown of insulation or refactory lining).
Checking furnace and tube temperatures.
Facing brick and refactory temperatures.
'Trouble shooting' on machinery, duct work, bearings, service mains.
Electrical wiring, motors, transformers, etc.
Checking process systems.
Balancing heating elements.
Temperatures of molten metals.
Ascertaining temperature profiles.

Leak detector

A simple, hand-held, battery-operated instrument (Figs. 4.5 and 4.6) designed to detect rapidly minute air or gas leaks in both pressure and vacuum systems. Any leaks which can be detected by water-bath or soap solution techniques can be more quickly and conveniently located by this ultrasonic detector. Visual indication is provided on a meter and audible indication is by means of detachable headphones.

Sensitivity: Will detect a leak through an orifice of 0·25 mm dia. (0·01 in) at a pressure of less than 0·7 kg/cm² (10 lb/in²) at 14 m (45 ft) range, or 0·05 mm dia. (0·002 in) at a pressure of less than 0·14 kg/cm² (2 lb/in²).

Industrial stethoscopes

A Tunable Stethoscope is a highly sensitive listening device. Faint noises are highly amplified for easy location. Unwanted noises can be reduced by a tuning control to pinpoint trouble spots. The instrument is of value in the preventive maintenance of all types of mechanical and electrical equipment by locating quickly and accurately wear and other

4.5 Ultrasonic leak detector being used with a contact probe to detect internal leaks in the pneumatic braking system of road petrol tankers

defects in running machinery. Trouble spots such as broken ball races and worn bearings can be pinpointed and rectified before a major breakdown occurs, thus saving costly damage repairs and reducing machinery downtime.

The instrument may also be used for tracing fluid flow in pipes, and the source of noises transmitted through solid objects and structures. Live electrical equipment may be examined safely using an insulated probe.

Stroboscopes

In stroboscopic light, rotating, reciprocating and vibrating mechanisms appear to be stationary or made to go through their operations in slow motion (Fig. 4.7). This enables moving parts to be studied while the plant is in operation – faults and defects can be observed which might not be visible were the parts stationary.

4.6 Ultrasonic leak detector used to detect external leaks

Vibration meters

During operation, rotary machinery generates vibrations. Sensors placed on stationary and moving parts of the machine (Fig. 4.8) are able to pick up and measure the various characteristics of these vibrations. Should the vibrations deviate from that which has been established as normal then this is a warning of impending trouble. By using a vibrations analyser, the precise reason for this change can be determined. This method of checking and monitoring enables the internal condition of the machine to be assessed accurately. Even a hair line crack in a turbine blade is reported to have produced a change in the vibration level.

Knowing the machine's condition:

(a) enables overhauls to be planned *when* they are required, and
(b) the possibility of unexpected breakdown is minimized.

Dye penetrants (for crack detection)

Dye penetrants provide a simple, cheap but also a sensitive searching means of detecting the existence of surface cracks and faults in manufactured components. The component under examination is degreased and the dye applied. A short period of time must elapse to ensure complete penetration and then excess dye is washed off, the component is dried and the developer applied. Any flaws present will be revealed through the developer. If fluorescent dyes are used, an ultra violet light is necessary to observe the flaws. Both the dye and the developer are obtainable in aerosol form.

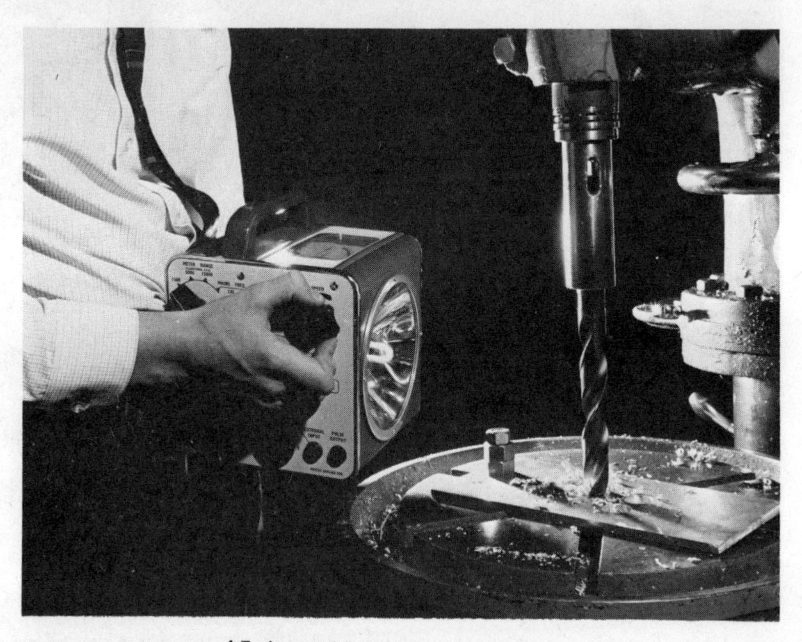

4.7 A portable transistor stroboflash

Ultrasonic flaw detectors

Portable, battery powered, ultrasonic flaw detectors (Fig. 4.9) are ideal for on-site, non-destructive testing. They provide quick and accurate location of structural or physical defects and thickness gauging on most materials.

The instrument uses the pulse-echo technique. An ultrasonic pulse is transmitted into the work piece by a transducer which is pressed against the face of the material. The pulse travels through the material and is reflected upon striking a flaw or another face. The time interval between the pulse leaving the transducer and the reflected pulse being received at

4.8 A portable transistor vibration meter in use

4.9 An ultrasonic flaw detector

the transducer is a function of the thickness or depth of the flaw. This time interval is measured and converted into a wave form which is projected onto the screen of a calibrated cathode-ray tube. Range: 5 mm (0·2 in) to 15 m (600 in) in steel.

Wall thickness gauge

The D-Meter Wall Thickness Gauge (Fig. 4.10) is a small portable digital display ultrasonic unit designed for making rapid non-destructive thickness measurements on steel or aluminium (modification for other materials is possible). As it requires access to only one side of the material being measured it is ideal for site work. Plant need not be taken out of service while measurements are being made even if the temperature of the measured surface is as low as −10 °C or as high as 500 °C (+14 °F to 930 °F). Special operator training is not required.

4.10 A D-meter used to gauge the thickness of pipes

The instrument will measure thicknesses from 1·2 mm to 300 mm (0·048 in to 12 in) with an accuracy of ±0·1 mm (0·004 in) over its whole range.

Application: Universal wall thickness gauging.

Thickness of ship's plate; extent of eroded or corroded areas thickness of pipe walls; erosion at bends. Depth of erosion or pitting. Wall thickness of pressure vessels, petroleum and other storage tanks.

Optical inspection devices

Bore viewers. These are simple, robust instruments for viewing the inside of small bores, tubes and holes. They can be used to inspect bores with diameters of 5 to 32 mm ($\frac{3}{16}$ to $1\frac{1}{4}$ in) and up to a length of 200 mm (8 in).

An Allen Bore Viewer (Fig. 4.11) consists of a small light probe and magnifier. At the end of the probe is a miniature lamp powered from the battery handle. A separate sleeve with a 45° mirror fits over the light probe.

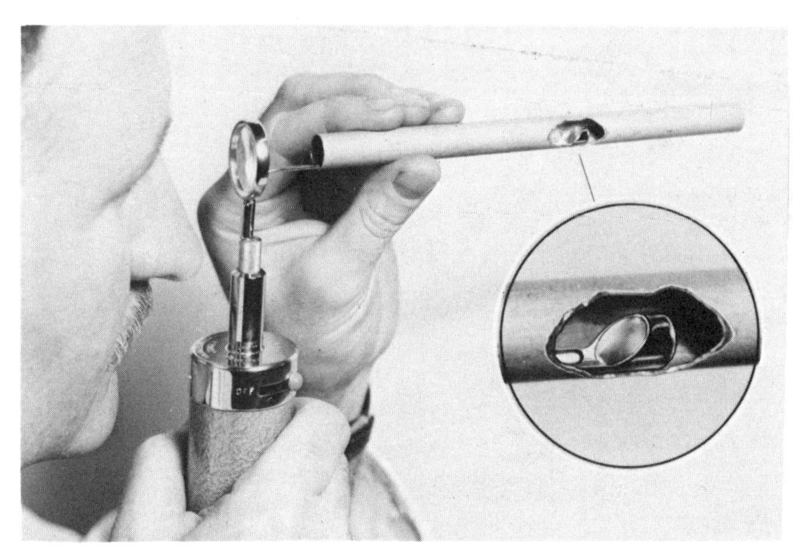

4.11 A bore viewer

Tank viewers. The Allen Tank Viewer (Fig. 4.12) is a portable optical instrument which is used to inspect the internal surfaces of medium sized tanks, vessels, etc., through a small opening – minimum 50 mm (2 in) diameter.

The instrument is 920 mm (36 in) long × 44 mm ($1\frac{3}{4}$ in) in diameter, at the outer end of which is a scanning prism and a low voltage 18 watt lamp.

Endoscopes. Endoscopes are slim tubular optical instruments that enable the user to look inside cylinders, tubes and similar hollow parts, particularly when the access to the hole is small.

The diameters of endoscopes (Figs. 4.13a to 4.13d) range from 5 to 45 mm ($\frac{7}{32}$ to $1\frac{7}{8}$ in) with length varying from 100 mm to 10 m (4 in to 30 ft), according to diameter. Endoscopes over 1 to 1·2 m in length are normally made in sections for ease of handling and transport.

Inspection mirrors. Inspection mirrors (Figs. 4.14a to 4.14c) are used for the internal inspection of larger hollow components and for viewing normally inaccessible parts of all kinds of plant, machines and equipment.

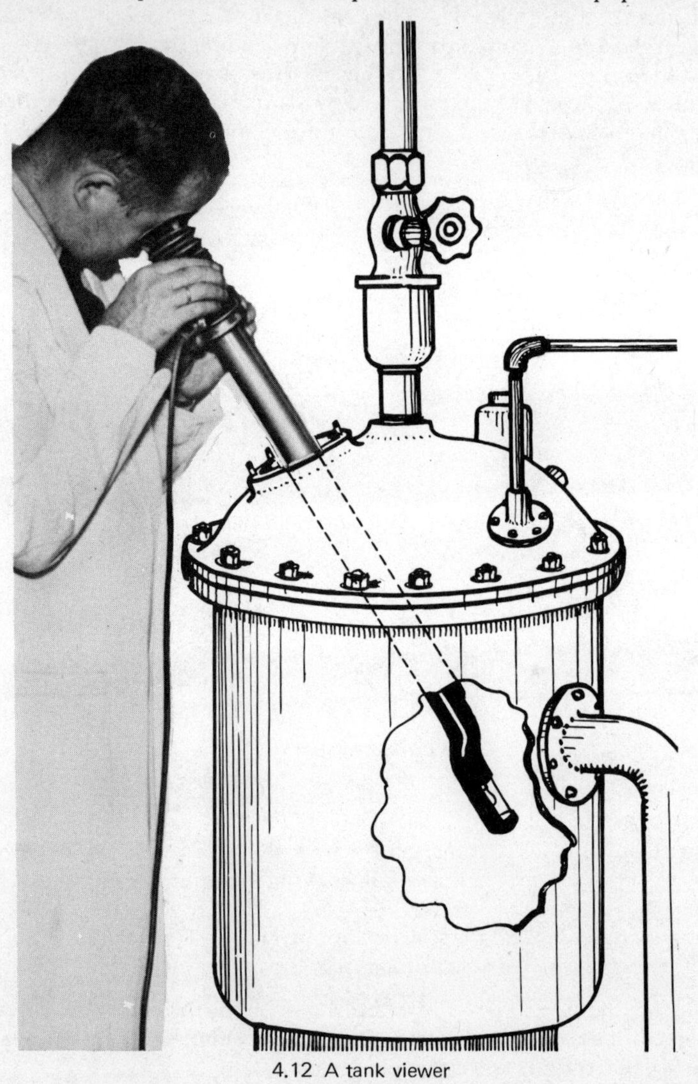

4.12 A tank viewer

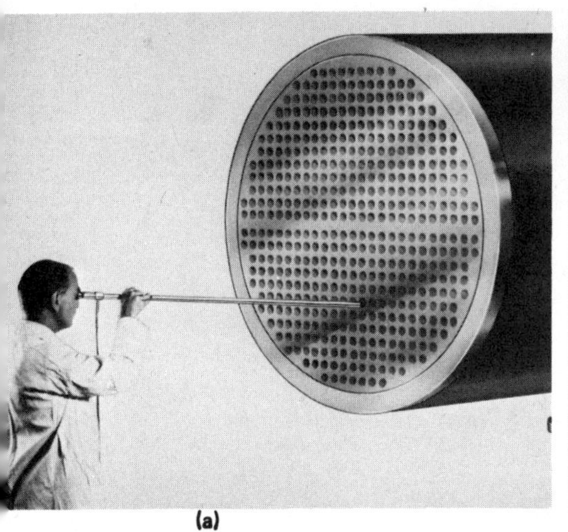

(a)

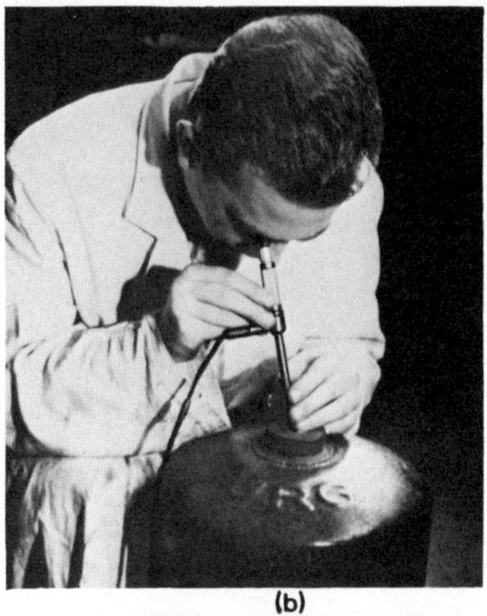

(b)

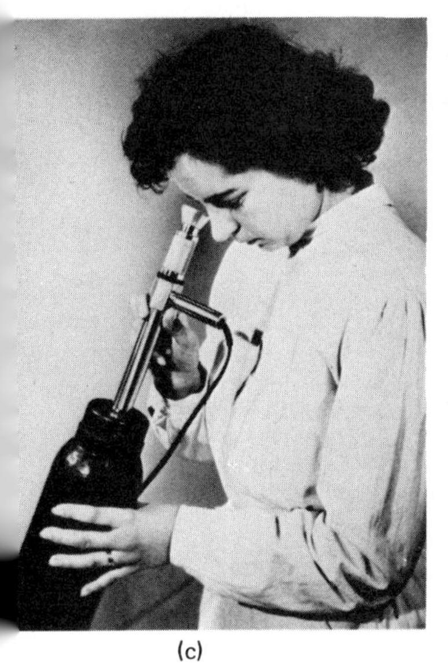

(c)

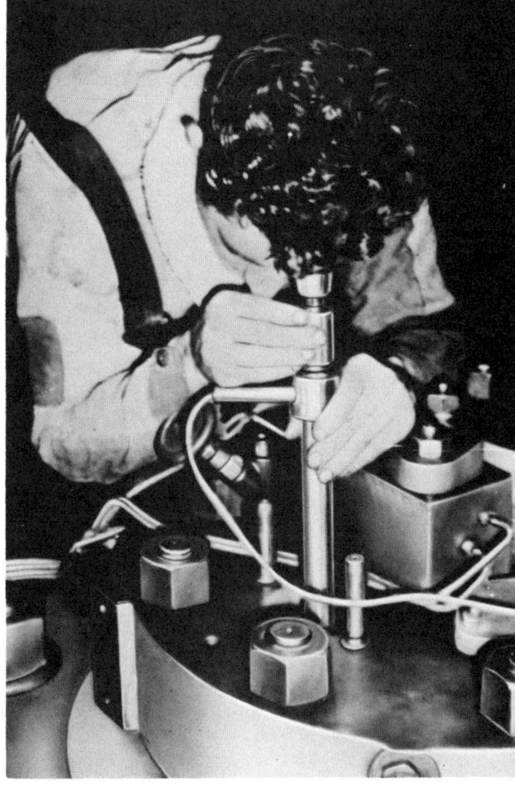

(d)

4.13 Endoscopes
(a) Inspecting a heat-exchanger
(b) Inspecting welds in gas tanks
(c) HP gas cylinder inspection
(d) Inspecting inside a large marine diesel engine cylinder without dismantling

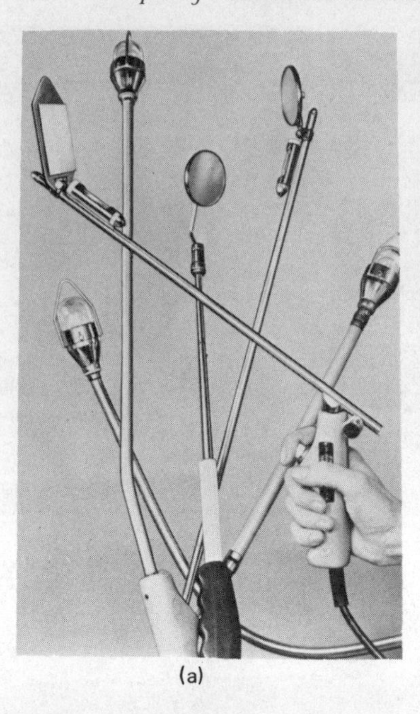

(a)

(b)

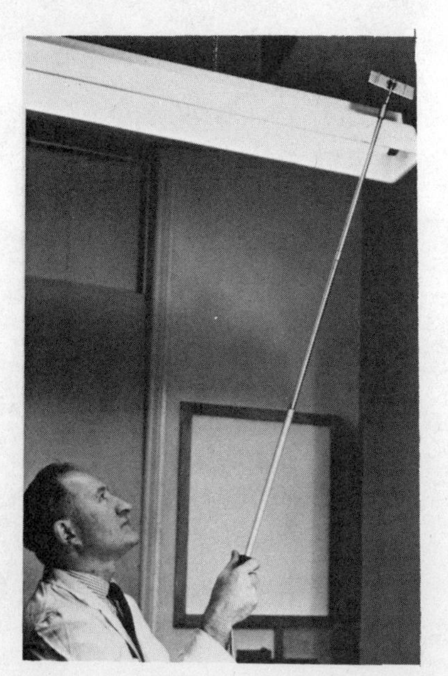

(c)

4.14 Inspection mirrors
(a) Angle of mirror adjustable by enclosed mechanism, operating control in handle
(b) Illuminated inspection mirror, inspecting a fire extinguisher
(c) Telescopic inspection mirror for viewing inaccessible places

Fibrescopes. Fibrescopes are portable flexible inspection instruments for the internal inspection of curved pipes, shaped hollow components and machinery (Fig. 4.15).

They consist of a vinyl-covered, fully flexible metal tube containing two glass fibre bundles, one for image transmission and the other for lighting. There are optical systems at both ends of the image bundle, a variable focus objective system at the distal end, and a focusing magnifier at the eyepiece end. The remote focus control for the objective lens focus is next to the eyepiece.

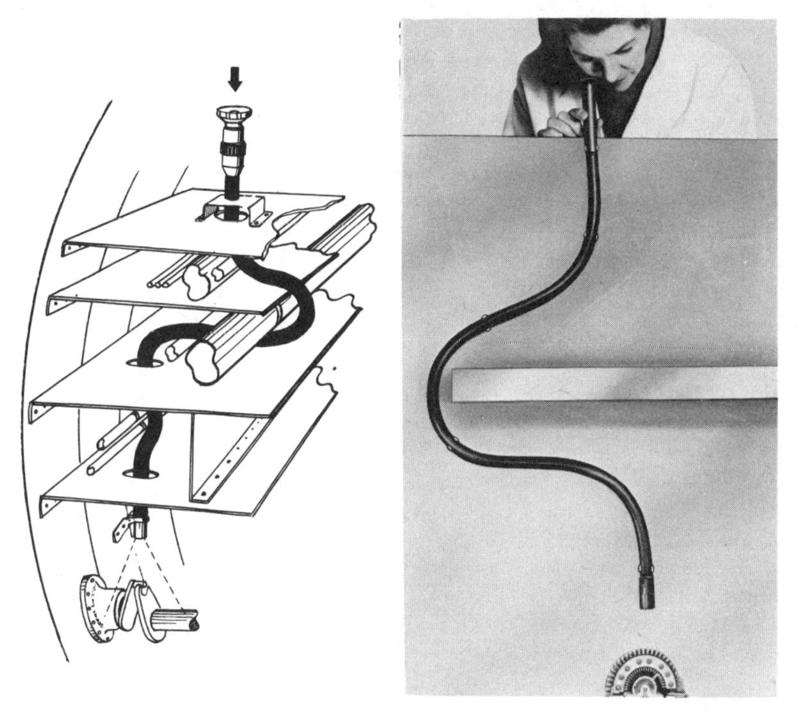

4.15 A permanently mounted flexible fibre-optic periscope which has applications ranging from in-flight undercarriage inspection to routine checking of inaccessible moving parts

Light from a high power light box is 'piped' by a flexible light cable to a connector on the internal light guide near the eyepiece. The transmitted image is broken up into many thousands of separate components, one for each fibre. The resultant picture therefore appears very like a high quality printed half-tone illustration.

The exceptionally high quality image produced by Fibrescopes is due to the precision clad-glass fibres used and the meticulous attention paid in manufacture to lay, pack and end finish.

Focus range: Adjustable from a few millimetres to infinity. Owing to the relatively small sized image formed on the bundle face, Fibrescopes are only intended for viewing in detail parts that are close – 250 mm (10 in) or less.

Focus control: Movement of the objective lens is controlled by a knurled sleeve at the eyepiece.

4.3. Charts and graphs

There are numerous types of charts and graphs used in industry, almost without exception their object is to illustrate the relationship between various functions. They are a visual aid in the process of communication, but to be acceptable as such they must be as easily understood as the written word or sets of tabulated figures, and the principle points under review, its message, must be clear at a glance.

Most students will be familiar with the construction, presentation, and use of the more general types of graph so these will not be discussed further. There are, however, other types of a more specialist nature which can be employed usefully by the maintenance department but with which students may not be so well acquainted. Three types of chart, although employed widely in many branches of industry, will be considered only from the point of view of assisting the maintenance function.

The Gantt chart

The Gantt chart is a form of bar chart originated by the American engineer Henry L. Gantt (1861–1919) which, since its introduction to industry (about 1915), has proved a valuable aid to planning and controlling a wide range of operations, including plant overhauls and installation projects.

The chart is constructed on a rectangular basis, all activities and tasks making up the project are listed separately down the vertical axis on the left hand side, while the time flows from left to right along the horizontal axis. The unit of time can be of any duration, but for most practical purposes either days or weeks are used.

Against each task is drawn a pair of horizontal parallel bars, the length of the lower bar represents the planned duration of the task, while the length of the upper bar is proportional to the amount of work actually completed at a particular time. This latter bar is adjusted in length as the work on the particular task progresses so that it can be compared with the lower bar (Fig. 4.16a).

Alternatively, instead of two parallel bars, the extremities of the lower bar only are indicated, the space between them is filled in as the work on the task proceeds. This latter adaptation saves vertical space and enables a longer list of tasks to be accommodated but it is not so distinctive when viewed from a distance (Fig. 4.16b).

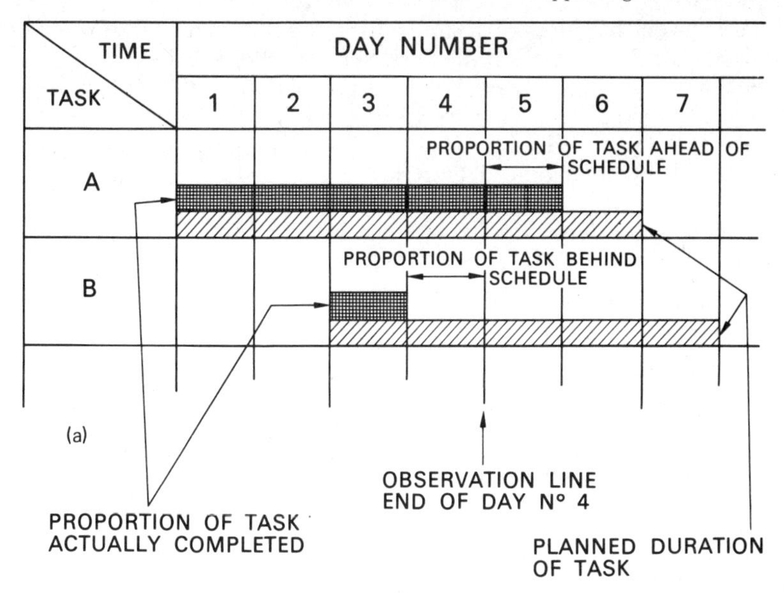

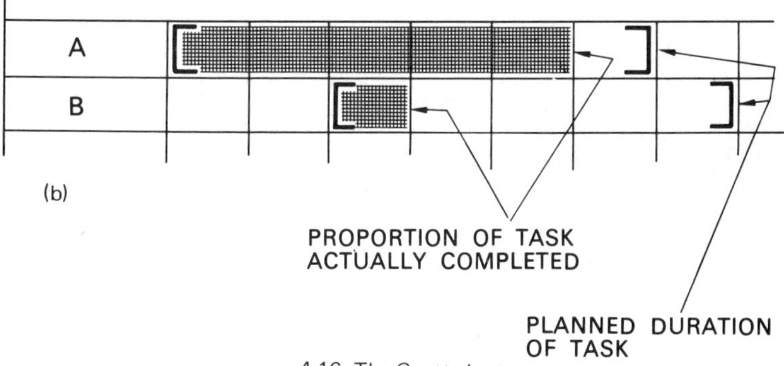

4.16 The Gantt chart

Whichever method is used, the bar representing the task's planned duration is positioned in line with its respective task and located along the horizontal axis between the task's planned commencing and completion dates. As the task proceeds, so the length of the upper bar or portion filled in between the extremities is increased to represent the amount of work completed – if 50 per cent of the task is complete, then the upper bar will be half the length of the lower bar.

To ascertain the progress of the work and compare it with the programme, the vertical observation line is dropped down the column under the appropriate date. The difference between the observation line and the work bar (upper bar) to the left of it indicates the proportion of the task behind schedule, while work bars that extend to the right beyond the observation line indicate the proportion of the task ahead of schedule.

With the aid of colours, codes and symbols, much additional information can be included on the chart – trades, tradesmen, labour requirement and man hours, can all be indicated quite easily. Proprietary boards and accessories are available that will accommodate all these data, and similar boards show machine loading, plant utilization, labour utilization, etc.

Example 1

To illustrate the application of a Gantt chart, assume the work schedule for a plant overhaul to be as that tabulated below.

Activity	To be Started Week Commencing No.	Activity Duration (weeks)	To be Completed Week Ending No.
A	1	4	4
B	5	6	10
C	11	5	15
D	3	8	10
E	7	6	12
F	1	5	5
G	1	8	8

As the work proceeds so it is recorded on the chart, review its progress at the end of Week No. 7.

From the chart (Fig. 4.17) the following information is apparent.

Activity	Planned Progress of Task	Actual Progress of Task	Remarks
A	Should be completed	It is complete	
B	Should be $\frac{1}{2}$ completed	It is only $\frac{1}{6}$ completed	Behind schedule
C	Should not yet be started	Not yet started	
D	Should be $\frac{5}{8}$ completed	It is only $\frac{1}{2}$ completed	Behind schedule
E	Should be $\frac{1}{6}$ completed	It is $\frac{1}{2}$ completed	Ahead of schedule
F	Should be completed	Not yet started	Behind schedule
G	Should be $\frac{7}{8}$ completed	Fully completed	Ahead of schedule

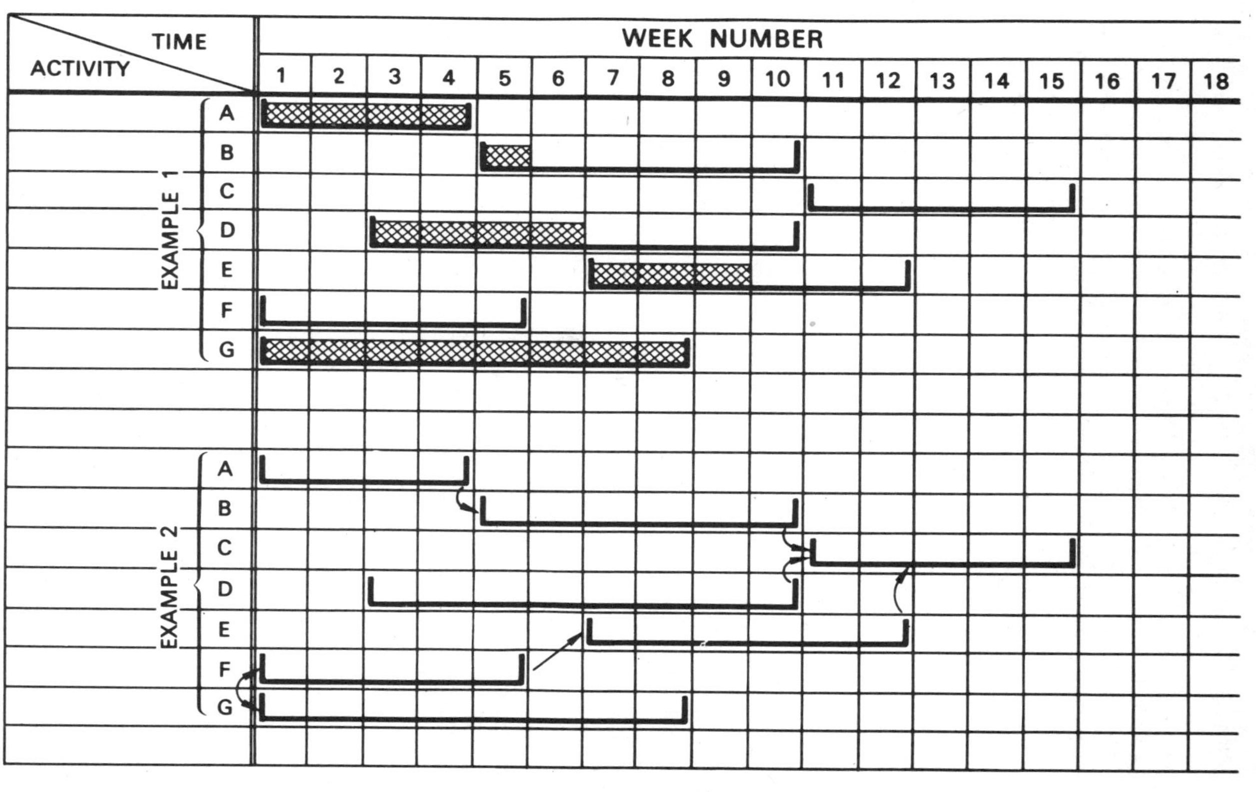

4.17 Application of the Gantt chart

Useful and informative as the chart may be, there is one aspect with which it cannot cope in its present form. It does not indicate clearly the interdependence of the various activities upon each other. Consequently, it is unable to identify or differentiate those particular activities that must adhere strictly to the schedule if the overall project is to be completed on time, and those activities in which some latitude is acceptable. To help overcome this limitation it is possible to apply the use of *links* to connect the various inter-related activities. The best way to illustrate their use is shown in Example 2.

Example 2

The activities, durations, starting and finishing dates are the same as those given in Example 1, but in the case of Example 2 connecting *links* have been applied to the chart. Observe the additional information that can now be expressed by their use (Fig. 4.17).

Activity A must be completed before Activity B can start.

Activity B must be completed before Activity C can start.

Activity D must be completed before Activity C can start.

Activity E must be completed at a point when Activity C is two-fifths completed.

Activity F must be completed before Activity E can start, but in this case there is 1 week to spare between the finish of F and the start of E so the completion of F is not so critical as the completion of A, B, D and E.

Activity F and G must start together.

The completion of Activity G is not decisive so long as it does not extend beyond the project completion date, i.e. week ending 15.

Activities A, B and C follow each other directly and together form a continuous chain of activity from the very beginning to the very end of the project. Thus, rigid adherence to the scheduled starting and finishing time of these three activities is crucial as there is no spare time (*float* in critical path method terminology) that can be used to prevent the planned project time being exceeded, should one of these activities fall behind schedule. In a critical path network A, B and C would be critical activities and form the critical path. All the other activities (D, E, F and G) are free to float, either at the start or finish, within their own specific confines. This latitude provides a limited manoeuvrability to meet the overall target date, even if their progress is not up to schedule or their timing and duration is slightly modified.

If we have to reduce the overall programme time, then the critical activities (A, B and C) must be the first to receive attention; the durations of these are shortened until the required time span is achieved. But by doing so, the float may disappear from other activities so they too in turn become critical and must also receive special attention.

Although the activities and durations are the same in both Example 1 and Example 2, the inclusion of links in the latter example increases greatly the chart's potential and value to the person planning or controlling the project. It highlights those tasks over which a very tight control must be exercised, and shows the overall effect of work which is either accelerated or delayed.

Ideal as the chart is for small or medium size projects, it can become rather confusing on large, complex operations. In such cases, the use of the critical path method of planning is recommended and in present day operations is the one that is generally used.

'Z' charts

Management is required continually to make decisions upon which subsequent actions are based. A particular course of action, once adopted, has then to be assessed. In many cases, the results may not be apparent for a considerable length of time, the effect taking place very gradually. Certain other actions may result in adverse short term effects which will be more than compensated in the long run, and many actions taken in one financial or calendar year do not show results until the next.

The full extent of these changing conditions are not always appreciated – factual comparisons may be difficult to make or confirm in the absence of long-term continuous records. Nor is the trend of a change easy to detect and assess if it is taking place simultaneously with operational and/or seasonal fluctuations.

With the passage of time, recollections become blurred, the human memory cannot be relied upon to recall unaided even a limited amount of information on past conditions with any degree of certainty. Thus, for management to judge the results of its efforts a simple recording means is required that will compensate for operational and seasonal influences, so a more balanced comparison can be made of past and present performances.

As an example, consider a continuous process plant whose total annual downtime, including that for maintenance, is budgetted carefully. As downtime forms an important part of the production costs, it has to be contained within its prescribed limits. For traditional as well as for statutory purposes, budgeting and the final accounting are carried out usually on an annual basis, but the operational cycle of many continuous process plants is planned on a much longer time span. It is not uncommon for an extended plant shut-down, for overhaul and modification, to take place only once in every two or three years.

Therefore, in order to accommodate both the financial and operational statistical requirements and also to show their relationship in the correct perspective, records must clearly indicate the annual results against those covering the complete operational time cycle.

Where results or information can be measured statistically the 'Z' Chart (so named because the curves of the completed chart form the letter Z) is able to fulfil this dual purpose as it provides a continuous graphical record of the same variable in three different forms:

(a) Current weekly or monthly results.
The actual current achievement is charted regularly. The short term effects of operational and seasonal variations are indicated clearly.

(b) Cumulative total of weekly or monthly results.
As the financial or calendar year proceeds weekly or monthly, comparisons can be made with both the position in previous years and the budgetted position of the current year.

(c) Moving total of results.
Operational and seasonal fluctuations are smoothed out so the general trend becomes clearer.
Note: The period of time (1 year, 2 years, 3 years, etc.) over which the total is taken will depend upon the particular situation, but it must be of sufficient length to encompass the complete operational or financial cycle, whichever is the longer.

Figure 4.18 illustrates the general appearance of the completed chart which in this case spans a period of one year. The figures quoted could refer equally well to either the cost of maintenance for a particular item of equipment or the downtime hours of a production process.

The moving total (in this instance, the moving *annual* total) is the sum of the monthly results of the last 12 months. When the results of the current month become available, they are added to the present total and the result of the corresponding month last year is then deducted. The moving total varies each time, either up or down, by the difference between the amount added and that subtracted.

The cumulative total is the accumulated total of the monthly results for a particular calendar or financial year only. The curve indicates the progress of performance only from the beginning of that specific year to the end of that year, or to date in the case of the current year. There is no carry-over of results from one year to another.

When we need a continuous long term picture of trends and fluctuations it is relatively simple to extend both the moving total and current monthly curves while still showing the cumulative curves for their respective years (Fig. 4.19).

Where the financial and/or operational cycle spans only one year then a moving *annual* total is sufficient to produce a curve of the required form. However, where the cycle time is two or even three years with perhaps high peaks of the charted function occurring irregularly throughout the cycle, then a moving annual total will not suffice. The curve formed in this manner will be very erratic and provide no useful information. To overcome this drawback, the data must be charted as a moving *biennial* total (two years span), or moving *triennial* total (three years span), whichever fits the case in question.

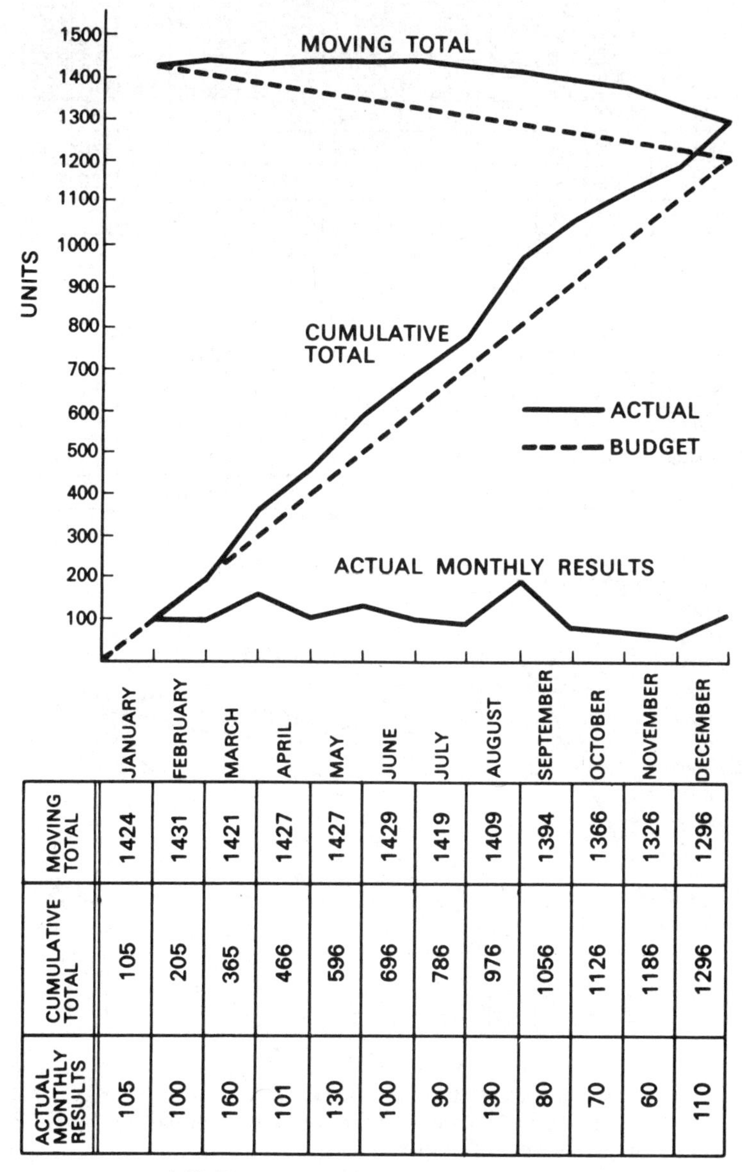

	MOVING TOTAL	CUMULATIVE TOTAL	ACTUAL MONTHLY RESULTS
JANUARY	1424	105	105
FEBRUARY	1431	205	100
MARCH	1421	365	160
APRIL	1427	466	101
MAY	1427	596	130
JUNE	1429	696	100
JULY	1419	786	90
AUGUST	1409	976	190
SEPTEMBER	1394	1056	80
OCTOBER	1366	1126	70
NOVEMBER	1326	1186	60
DECEMBER	1296	1296	110

4.18 The general appearance of a completed Z chart

	MONTHLY RESULTS	CUMULATIVE TOTAL	MOVING ANNUAL TOTAL	MONTHLY RESULTS	CUMULATIVE TOTAL	MOVING ANNUAL TOTAL	MONTHLY RESULTS	CUMULATIVE TOTAL	MOVING ANNUAL TOTAL	MONTHLY RESULTS	CUMULATIVE TOTAL	MOVING ANNUAL TOTAL
JANUARY	105	105	1424	50	50	1241	55	55	800	60	60	936
FEBRUARY	100	205	1431	52	102	1193	58	113	806	65	125	942
MARCH	160	365	1421	100	202	1133	120	233	826	125	250	947
APRIL	101	466	1427	55	257	1087	60	293	831			
MAY	130	596	1427	72	329	1029	80	373	839			
JUNE	100	696	1429	45	374	974	58	431	852			
JULY	90	786	1419	55	429	939	70	501	867			
AUGUST	190	976	1409	120	549	869	135	636	882			
SEPTEMBER	80	1056	1394	60	609	849	70	706	892			
OCTOBER	70	1126	1366	50	659	829	65	775	907			
NOVEMBER	60	1186	1326	56	715	825	70	845	921			
DECEMBER	110	1296	1296	80	795	795	90	935	931			

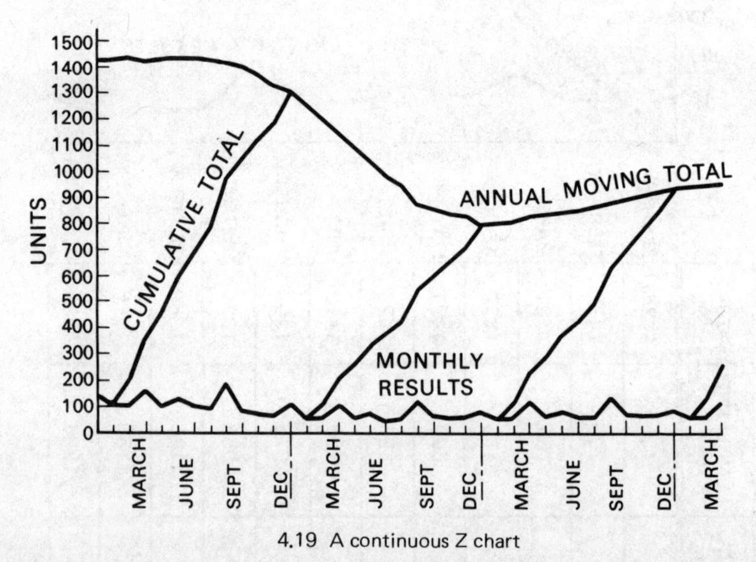

4.19 A continuous Z chart

Note: To calculate the moving biennial total take the sum of 24 consecutive months. To calculate the moving triennial total take the sum of 36 consecutive months and apply in a similar manner to that described for the calculation of the moving annual total.

The tabulated data set out in Fig. 4.20 represent the downtime hours of a continuous process plant over a period of four years. It is inferred from

	1967				1968				1969				1970			
	Monthly Results	Cumulative Total	Moving Annual Total	Moving Biennial Total	Monthly Results	Cumulative Total	Moving Annual Total	Moving Biennial Total	Monthly Results	Cumulative Total	Moving Annual Total	Moving Biennial Total	Monthly Results	Cumulative Total	Moving Annual Total	Moving Biennial Total
January	102	102	1406	2565	105	105	1188	2594	50	50	1198	2386	55	55	642	1840
February	93	195	1404	2568	100	205	1195	2599	52	102	1150	2345	58	113	648	1798
March	110	305	1424	2583	95	300	1180	2604	55	157	1110	2290	60	173	653	1763
April	95	400	1421	2573	102	402	1187	2608	55	212	1063	2250	57	230	655	1718
May	91	491	1412	2562	103	505	1199	2611	48	260	1008	2207	52	282	659	1667
June	98	589	1407	2567	100	605	1201	2608	45	305	953	2154	58	340	672	1625
July	100	689	1402	2569	90	695	1191	2593	55	360	918	2109	62	402	679	1597
August	98	787	1170	2567	340	1035	1433	2603	58	418	636	2069	310	712	931	1567
September	95	882	1175	2570	50	1085	1388	2563	60	478	646	2034	50	762	921	1567
October	98	980	1175	2571	50	1135	1340	2515	50	528	646	1986	55	817	926	1572
November	100	1080	1180	2568	60	1195	1300	2480	56	584	642	1942	54	871	924	1566
December	105	1185	1185	2582	58	1253	1253	2438	53	637	637	1890	53	924	924	1561

4.20 Data from a continuous process plant

the figures that every two years, in August, the plant is shut down for approximately two weeks, during which time overhaul and modifications are carried out. It would appear that as a result of the 1968 overhaul the plant performance was considerably improved, indicated by the drastic reduction in downtime.

Exercise 1

From the given information (on page 131) construct a 'Z' chart by plotting the following curves:

1. The monthly results continuously over the four year period.
2. The cumulative total in respect of each individual calendar year.
3. The moving total continuously over the four year period:
   (a) as a moving biennial total*, and
   (b) as a moving annual total.

Give an interpretation of the curves produced.

Pareto curve

During the course of his work, Vilfredo Pareto (1848–1923) an Italian economist and sociologist, analysed the distribution of wealth. He concluded that most of it was controlled by a minority (the vital few), while the remainder of it was distributed among the vast majority (the trivial many).

The principle of Pareto's Law of Maldistribution can be applied to analyse many managerial situations and problems, provided that each of the contributing elements can be expressed in measurable terms. It assumes that there is a small number (the vital few) of dominant elements which exert the greatest influence upon the ultimate overall result, while the effect of a large number (the trivial many) of the elements is minimal. This can be represented graphically by plotting the elements concerned against their contribution to the total effect. A curve (Pareto Curve) is produced which provides a means of identifying the vital few and the trivial many, thus forming a basis for selective or differential control. The general pattern of the curve is similar to that shewn in Fig. 4.21a. The curve can be divided into three distinct sections, A, B and C; for this reason it is sometimes known as an A-B-C curve. Section A identifies the few elements which exercise the greatest influence while Section C identifies the elements which contribute least to the overall result. Occasionally the curve may be considerably flatter and the distinctive A B C

*Note: In this particular example there is a two year operational cycle, therefore, the moving biennial total is the most suitable to apply in the circumstances.

The moving annual total has been included to illustrate the effect described in the text.

In order to show both the monthly results and also the cumulative total in sufficient detail, it may be necessary to plot on the same chart the moving biennial (or triennial) total to a different scale.

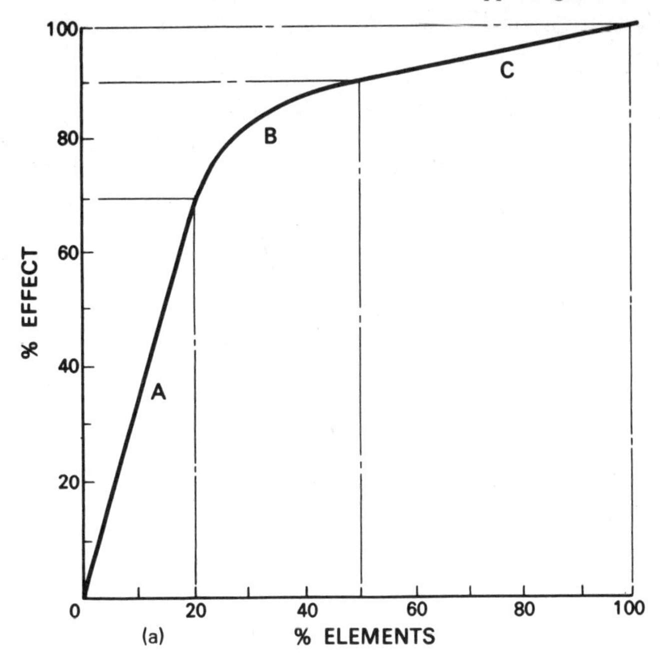

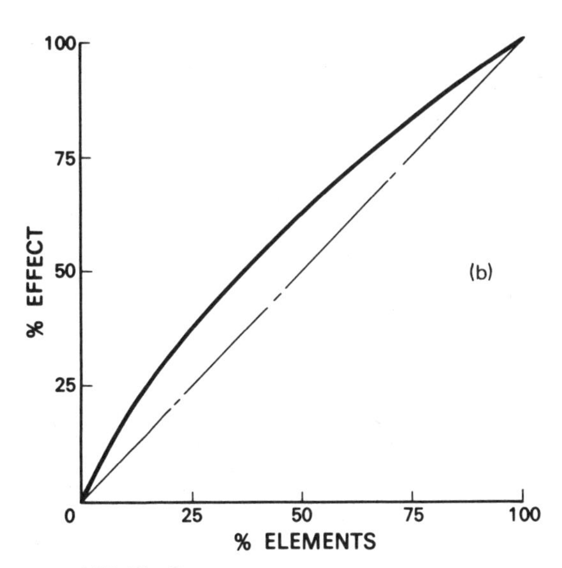

4.21 The Pareto curve
(a) The general pattern (b) The flatter curve

characteristics become difficult to distinguish. The nearer the curve approaches a straight line the more evenly does each element contribute to the overall result (Fig. 4.21b).

Referring to the particular curve illustrated in Fig. 4.21a, only 20 per cent of the total elements are in Section A but they are responsible for 70 per cent of the ultimate result. Section B contains 30 per cent of the elements and these influence the results by 20 per cent. The remaining 50 per cent of the elements which form Section C have very little bearing upon the ultimate result and affect it by only 10 per cent. It follows that by exercising a tight control or close scrutiny over the few elements forming Section A particular attention is being directed to 70 per cent of the overall result. Hence, the greatest attention is focussed upon a comparatively small sector where it will be of most use. Less stringent forms of control are required for elements in Sections B and C, as their overall effect, particularly those of Section C, is of a much lower order. A small percentage change in the value of the elements forming Section A is usually much easier to achieve and has a much greater effect than a large percentage change in the value of the elements forming Section C.

However, not every analysis is concerned primarily with the identification of the vital few; the crux of many problems lies in the determination of the elements that contribute little to the ultimate effect. It is the elements forming Section C that must now receive the closest attention, the inference being that as the overall contribution of these elements is relatively small their complete elimination will have little effect upon the ultimate result. Consequently, the resources that are now expended upon them could be put to better use elsewhere. Alternatively, steps could be taken to improve their contribution to a more acceptable level.

The practical value of this form of analysis lies in its ability to provide management with a clear, diagrammatic arrangement showing the relative importance of the factors involved. Consider a situation consisting of 20 different elements with their respective values tabulated below.

Element	Elemental Value	Element	Elemental Value
A	302	L	353
B	52	M	187
C	13	N	13
D	12	P	35
E	360	Q	48
F	150	R	348
G	350	S	18
H	26	T	39
J	76	U	74
K	19	V	25

In this example, the units of value have not been defined but in practice they could represent:

Money – £
Time – hours
Units of production
Units of consumption etc.,

depending upon the type of situation being analysed.

Method of construction:

1. Determine the elements that form the situation and express each element in measurable terms (see opposite).
2. Construct a table with five vertical columns each headed as shown in Fig. 4.22.
3. List all the elements in descending order of value – column 1.
4. Against each element enter its respective individual value – column 2.
5. List the cumulative values of the elements – column 3.
6. Express each of the cumulative values as a percentage of the final cumulative value – column 4.

   The final cumulative value is 2500, this represents the total effect (i.e. 100 per cent) of the situation. Therefore, expressing a value as a percentage of this final cumulative value is the same as defining its proportional relationship (i.e. its contribution) to the total effect.
7. Working from top to bottom of the table, express each element, on a cumulative basis, as a percentage of the total number of elements. There are 20 elements so each one will account for 5 per cent of the total number – column 5.
8. Plot a curve using the figures shown in column 5 horizontally against the figures shown in column 4 vertically (Fig. 4.23). The methods of constructing the curve for the determination of the 'vital few' or the 'trivial many' are exactly the same. It is only the interpretation that is different.
9. By inspection, the curve can be marked off into its characteristic sections of A, B and C.
10. To establish the identity of the individual elements forming each section, cross reference should be made between the curve, column 5 and column 1.

This form of analysis can be applied to a variety of subjects. A few of the more common and obvious ones are listed below, from these others may be suggested which meet specific needs or circumstances.

Cost analysis – identification of high/low cost items, operations, processes or elements.

Time analysis – identification of high/low time consuming operations, processes or elements.

Col 1	Col 2	Col 3	Col 4.	Col 5	
Element	Elemental Value	Cumulative Value of Elements	Cumulative Value of Elements Expressed as a % of the Cumulative Total	% Elements	
E	360	360	14·4	5	
L	353	713	28·52	10	
G	350	1063	42·52	15	A
R	348	1411	56·44	20	
A	302	1713	68·52	25	
M	187	1900	76	30	
F	150	2050	82	35	
J	76	2126	85·04	40	
U	74	2200	88	45	B
B	52	2252	90·08	50	
Q	48	2300	92	55	
T	39	2339	93·56	60	
P	35	2374	94·96	65	
H	26	2400	96	70	
V	25	2425	97	75	
K	19	2444	97·76	80	C
S	18	2462	98·48	85	
C	13	2475	99	90	
N	13	2488	99·52	95	
D	12	2500	100	100	

4.22

Scrap or wastage analysis – identification of machines, operations or processes which produce the highest/lowest scrap rate.

Breakdown analysis – identification of machines operations or processes with the highest/lowest breakdown frequencies, times or costs. Causes of breakdown can also be analysed in this manner.

Stores analysis and control – identification of slow moving items.

The identification of items which account for either the highest/lowest useage, or represent the highest/lowest monetary value. By identifying the items into their relative sections, A, B or C, selective control can be exercised, i.e. a tight control over items appearing in Section A with progressively less or different control over items in sections B and C.

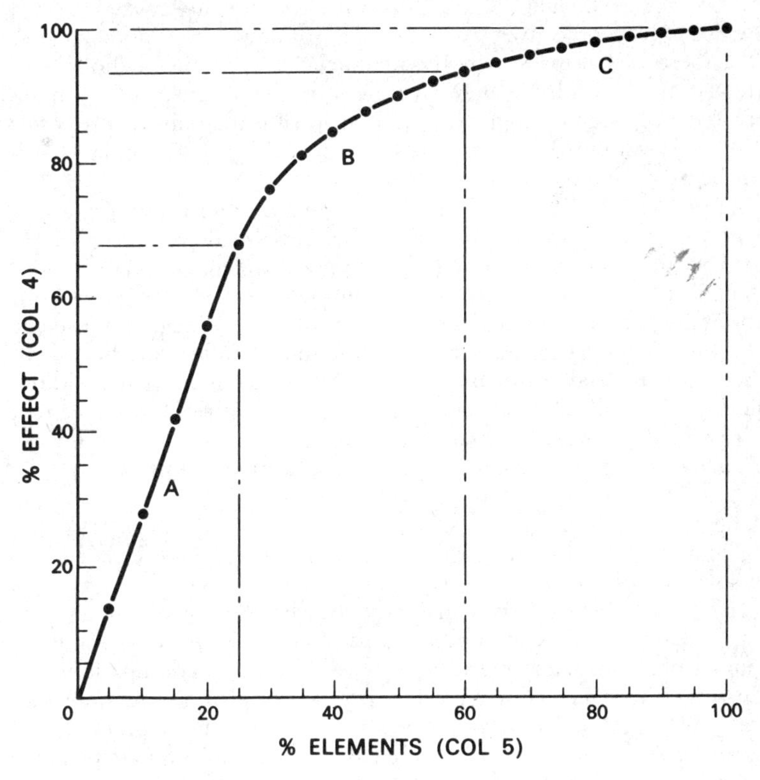

4.23 The completed Pareto curve

4.4. The human element

Manpower planning

The success or otherwise of any maintenance plan, no matter how well conceived, will be influenced directly by the performance, skills and attitudes of the staff. As much consideration should be given to the human element of the plan as is given to the more material elements.

Prior attention and forethought on the human side of a situation is within the general scope of manpower planning, although the extent to which it is practised varies enormously. Planning may be limited to the manning requirements of a future project or production process, while at the other end of the scale it involves all aspects of manpower requirements, recruitment, selection, promotion and integrates them into the overall company policy.

A realistic manpower plan offers a framework within which decisions and changes can be made with the assurance that they are always directed towards agreed objectives. In the absence of any corporate manpower plan, there is no reason why the maintenance department should not attempt to formulate their own within the context of planned maintenance. Even though it may not be possible to apply the proposals completely or immediately, the plan does provide a valuable guide which may be used should a suitable occasion arise.

Manpower requirements, selection, utilization and training are primary areas which are usually within the scope of departmental management and are a good practical basis upon which to initiate a plan. Each of these areas has a direct bearing upon the others – changing conditions and circumstances in any one affects the rest. The plan must take this close inter-relationship into account and be sufficiently flexible to accommodate, and compensate for, changes that may occur, without altering materially the ultimate objective. Irrespective of any apparent success of the plan, a periodic review is necessary to confirm its continuing validity with the company's latest policies and aims.

Manpower requirements

The objectives of the maintenance plan must now be translated into persons, trades and occupations.

From a detailed analysis of maintenance schedules, programmes, present and future plans, it should be possible to assess realistically the number of persons, and the various types of trades and occupations that constitute the *ideal* maintenance team to accomplish the maintenance work. We need an *ideal manpower standard* that is both practical *and* economical, and one that can be used as a measure of comparison and a target for the future. If the standard is to be of any real value, then transitional constraints such as personalities and inter-departmental politics should be ignored. Similarly, the tendency to over estimate manning requirement – to be on the safe side, to prepare for an emergency, to ensure labour availability at all times without undue pressure, etc., should also be curbed strictly.

An inventory of the present staff and their respective occupations can, when compared with the ideal standard, highlight areas of divergence. Under- or over-staffing can be noted for future action. Immediate or drastic action could create problems which outweigh the anticipated benefits, bring disrepute and foster resistance to future planning. However, knowing where the areas of improvement lie, we can work towards the ideal gradually. Vacancies created by the following often provide a suitable opportunity to change the balance of existing manning without too much upheaval:

(a) Natural wastage – retirements, persons leaving the company.
(b) Promotion – to another department or factory within the company, within the department.
(c) Expansion – of the company or department, construction of a new factory, extra shift working.
(d) Re-organization – introduction of new production or maintenance methods.

Before a vacancy is filled automatically by a person of similar occupation, reference to the standard and the inventory will indicate if and how the situation can be used to edge closer to the ideal.

The technical and personal qualifications of the person to fill a vacancy will vary not only from factory to factory but also from department to department. A small factory cannot usually afford to carry a large maintenance staff so a person may be called upon to undertake several diverse functions. The works electrician may carry out elementary instrument maintenance as well as his own particular trade. The maintenance fitter will be expected to have a working knowledge of and be able to maintain plant and equipment ranging from complex production machines to the factory central heating system. The shift foreman may not only be in charge of the production process but also be required to supervise and organize the shift maintenance. In such cases, the personnel of the maintenance department must be 'Jack of all trades' or at least be able to 'double up' in several capacities.

The situation is completely different in a large factory or industrial complex (i.e. paper mills, steel mills, oil refineries, coal mines, chemical plants), since there is usually sufficient work of a specialist nature to employ tradesmen solely in their own occupation. Rigid enforcement of this principle is often necessary because of demarcation and restrictive practices that tend to proliferate in large establishments. The maintenance team in this instance would comprise of mechanical fitters, electricians, instrument fitters, welders, builders, etc.

The staff is the company's most valuable asset, and like any other asset it is engaged to meet specific needs. These needs must be defined precisely, so that when filling a vacancy we know exactly what is required, and by careful selection get the right person. The first step in the selection procedure is to consider the vacancy and formulate a job description setting out the pertinent points. A properly drawn up description includes the essential duties of the job and the relevant supplementary details needed to complete the picture. A job described simply as works electrician, mechanical fitter or welder is no more explicit or descriptive than are the terms lathe, milling machine, grinder or drilling machine quoted on an order when purchasing machine tools. Some of the factors that build up the job description are listed below:

Job title – or name.

Location – factory, department or section in which the job is located.

Job summary – a brief summary of the job. What it involves, the relative times devoted to its various aspects. Degree of difficulty and/or complexity.

Responsibility – the job grade. The level of responsibility. The degree of supervision received.

Skill and ability – basic and specialist skills and/or knowledge required. The quality of the work and standard of accuracy demanded. Degree of intelligence, resourcefulness, judgement.

Technical and educational qualifications needed.

Physical nature – light, medium or heavy work. Indoors, out-of-doors, clean, dirty, physically demanding. Repetitive or varied. Sedentary or mobile: Suitable for handicapped persons. If height, enclosed spaces or colour identification is involved.

Social nature – does the work involve group or team work, or individuality.

Training – the training required to accomplish the job satisfactorily – the training given.

Conditions of employment – hours of employment – day work, shift work, overtime. Salary, fringe benefits, etc.

Promotional prospects – the opportunity for advancement within the section, department, factory and company that the job offers.

Having established the job's features, we should now consider the type of person best suited to perform it, and set out his desirable characteristics in the form of a personnel specification. Apart from the usual factors of age, experience, skill, etc., the job description should reflect any specific characteristics that are necessary, or in some cases, those that should be avoided. For example, if colour identification or confined spaces are involved, persons suffering from colour blindness or claustrophobia would be most unsuitable.

Although certain basic qualifications are necessary for all tradesmen, maintenance personnel need certain additional qualities that may not be quite so essential to persons engaged on routine production work. These are:

Self assurance – to confidently tackle any maintenance task that might arise.

Initiative – to act on own judgement when required.

Self reliance – able to carry out the work without constant supervision.

Stability of character – unflustered under pressure.

In a small factory, the selection and engagement of personnel will be carried out by the works manager, works engineer or departmental supervisor. To these persons, a job description and a personnel specification may seem superfluous; they are so near the job that they know in their own minds the details of the work to be done and the exact type of person they want to do it. It is, however, very useful to have at hand an aide memoir, perhaps in the form of a check list, to ensure that no detail is overlooked.

The situation is very different in a large factory where a central personnel department is responsible for all recruitment. Being more detached and not intimately concerned with the precise practical aspects of the work, a detailed job description and personnel specification are of vital importance. The selection of the right person from a number of candidates of similar calibre can prove very difficult. It thus becomes necessary to evaluate and present the qualities of each candidate in a form that will enable an overall comparison to be made with a common standard. The following method attempts this by numerically rating each candidate (see Fig. 4.24):

1. Compile a list of the qualities necessary for the job (list in descending order of importance).
2. Establish a rating scale for each quality.
3. Assess and rate the proficiency of each candidate for each quality, the total of each candidates marks will give a means of comparison.

This method should be used only as a guide to selection, usually in conjunction with an interview.

The efficient use of manpower

Many companies regard their maintenance department as an insurance policy – only to be used in cases of emergency. Some companies are quite prepared to allow their staff to wait around until called out, the inference being that if the staff are not working then the plant must be functioning alright. This attitude might have been acceptable in the past but the dictates of modern industry makes the efficient utilization of manpower an economic necessity.

A maintenance tradesman is employed because he possesses and practises specific skills. If he is to be utilized at his full capacity he must be occupied fully only on those tasks that demand his special expertise. However, 'most objective examinations of the situation, including activity sampling methods, have shown that maintenance men are usually working for only about one third of their paid hours. They spend another third walking about and the remainder of the time waiting for instructions, waiting to be served at the stores or waiting for the plant to be released by their production colleagues'[17]. But this analysis registers only the time the subjects were actually working, it has not taken into account the grade of work undertaken during this working period. If it is assumed, not unreasonably, that only 75 per cent of the work carried out demanded the specialist skills of a tradesman, the other 25 per cent could

PERSONNEL EVALUATION

Position:
Sectional Foreman – Maintenance Dept

PROFICIENCY CODE				
A	B	C	D	E
Excellent	Very Good	Good	Fair	Poor

Required Qualities ↓	Proficiency Rating Scale A	B	C	D	E	Candidate's Names J Brown	R Green	D White	K Black		Notes
Technical Skill and Ability	30	25	20	15	10	25	23	26	20		Minimum Marks Acceptable 22
Organizing Skill and Ability	20	16	12	8	4	15	16	12	14		
Leadership	20	16	12	8	4	12	13	10	14		
Resourcefulness	15	12	9	6	3	12	9	11	11		
Co-operation	15	12	9	6	3	11	11	12	11		
Temperament	15	12	9	6	3	11	10	11	9		Minimum Marks Acceptable 11
Max Marks Obtainable	115	Marks Obtained by Each Candidate				86	82	82	79		
REMARKS											

4.24 A personnel evaluation chart to assist in the selection of personnel for a specific position

have been done by less skilled persons, then the true or effective utilization of the tradesman is now only 25 per cent, i.e. $0{\cdot}33 \times 0{\cdot}75 = 0{\cdot}25 = 25$ per cent utilization. Thus, the effective utilization of tradesmen can be expressed as a product of:

(a) the proportion of time actually spent working, and
(b) the proportion of this time spent only on those tasks that demand the expertise of tradesmen.

Figure 4.25 illustrates the effect that manpower utilization has on productivity. A relatively small improvement in utilization has a considerable impact on productivity, particularly in the sector quoted in the above example. On these facts alone the need for, and the benefits of, better use of manpower resources is clear.

It is management's responsibility to ensure that the efforts of their staff are channelled in the right direction and utilized in the most productive manner. The misuse of manpower can manifest itself in different ways, some have already been mentioned, some may be controlled by local departmental management, simply by better planning. Initially, the survey to provide a basis for this planning should consider the tradesman's activities from two aspects:

1. Is the tradesman's *time* being used to the full?
   How much of the tradesman's time is actually spent carrying out maintenance work?
   How much time is spent –
     obtaining supplies and tools
     waiting for instruction
     travelling to and from the facility to be maintained
     waiting for the facility to be released by the production department?
2. Is the tradesman's *skill* being used to the full?
   Could some of the jobs or parts of them, at present carried out by a tradesman, be done by less skilled persons?
   Is the ability of each person utilized to the full?

A detailed analysis along the above lines will highlight areas where action can most profitably be taken.

Planned maintenance will improve the situation, but even so there will be times when a tradesman is waiting in the workshop 'on call'. On such occasions, he could be usefully employed on routine bench work that can easily be picked up and put down.

Although planning attempts to direct a person's activities, it cannot control his personal performance. A person's performance depends not only upon his ability but upon his willingness to apply that ability. In turn, the degree of willingness depends largely upon his personal attitude towards his work. A person's attitude reflects his state of mind towards his working environment (working conditions, supervision, personal relationships, company and departmental policies and practices). If this is

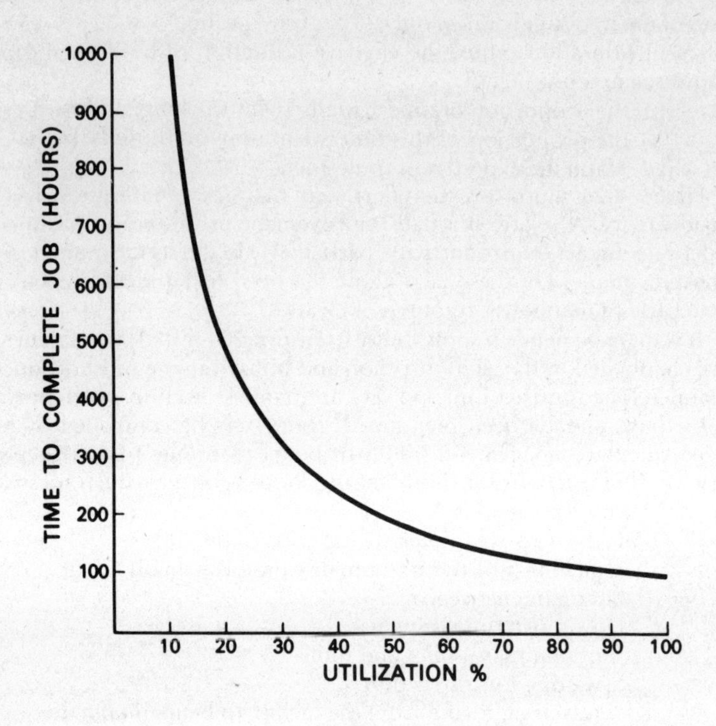

UTILIZATION %	TIME REQUIRED TO COMPLETE JOB
10	1000 hours
20	500 hours
30	333 hours
40	250 hours
50	200 hours
60	166 hours
70	143 hours
80	125 hours
90	111 hours
100	100 hours

4.25 Utilization-productivity curve

psychologically favourable to him he will give his best, if it is not, his work will suffer. Thus, it is in management's interest to ensure that the morale and enthusiasm of all its staff are maintained at a high level.

Training

In the past, when machines were relatively simple and change was a gradual, evolutionary process, a tradesman could assimilate new developments in the natural course of his everyday work. Basic skills learnt as an apprentice and consolidated by subsequent practical experience were usually sufficient to carry him through the whole of his working life.

Today this is no longer the case – machines are becoming more and more complex, new processes, ideas, methods, techniques and materials are being introduced faster than ever before. Advanced and sophisticated techniques which until a short while ago were associated only with scientific establishments now have widespread applications in industry. There has been greater technical advancement in the last decade than in the whole previous span of engineering, and there is no indication of it slowing down. To keep pace with these developments, maintenance is demanding the application of new skills and techniques, many of which are in areas previously considered beyond the scope of maintenance personnel. If the maintenance department is to fulfil its proper function in a progressive industrial society, then its personnel must be trained to meet current needs and future requirements.

Training should not be a 'once only' event but a continuous and progressive process designed to increase the individual potential of maintenance staff members and to form them into a technically qualified, well-organized, efficient team. Training need not always be of a purely technical or specialist nature; background subjects – maintenance techniques, maintenance organization, method study – that enable a person to understand the reason for and the purpose of his efforts, give an added dimension to his job. 'Know-how' together with 'know why' promotes awareness and stimulates interest, a person interested in his job will do it better.

The objective of a training programme must be defined clearly if it is to succeed and the areas in which it is most needed must be identified. Further examination within these areas will indicate the type of training necessary:

(a) to increase skill or technical knowledge, to introduce new techniques,
(b) to improve efficiency and morale,
(c) for promotion, to provide successors, and
(d) to increase potential.

Having decided upon the type of training required, we must consider how it is to be carried out. As so much depends upon local circumstances, the number and type of persons involved, the subject matter, and the level and depth of training, there is not one best way. Although training is most effective when the trainee participates actively in the learning process rather than simply listening to lectures, this should not be interpreted as working with an 'experienced' operator. This is far from ideal as bad habits and mal-practices are just passed on. The aim of training should be to teach the correct method from the start.

Training courses may be organized either on an internal or external basis.

Internal courses:

Organized by the company for its own employees, may cover such topics as company policy, processes, techniques, method study, labour relations, costs, etc., as applicable within the company's organization. Such courses may take the form of:

use of specialists from within the company to give specialist lectures or demonstrations, or

visiting lecturers or specialists.

Training within industry – TWI:

Arrangements with local colleges to run works based courses.

Courses held at the company's own training centre.

Lectures and demonstrations by manufacturers (demonstrating their own products).

Films and exhibitions, visual aids, posters, etc.

Visits to other departments or factories within the company.

External courses:

The vast majority will be those held at an educational establishment, the student (trainee) attending either

(a) a long term course in preparation for a nationally recognized academic or trade qualification, or

(b) a short course (usually of 1–2 weeks duration) which deals with a specific, often specialized, subject.

When the type and subject matter of the course have been established, the training needs of the persons concerned must be assessed so that a training programme can be devised. Figure 4.26 illustrates a training programme chart. It lists the name of each participant, together with the operations in which they must become proficient. The chart can be used:

(a) To assess the training needs of each participant. The person's proficiency in each particular operation or job is noted and compared with the minimum acceptable standard for that job (not all jobs require the same standard of proficiency). From this comparison, the training needed to reach the required standard can then be estimated.

Note: For convenience the degree of proficiency can be indicated by a code: A – Very good, B – Good, C – Fair, D – Poor, E – No previous experience.

TRAINING PROGRAMME

DEPARTMENT:– Maintenance

DATE OF ISSUE:– 5th. Feb. 1973

PROFICIENCY CODE				
A	B	C	D	E
VERY GOOD	GOOD	FAIR	POOR	NO PREVIOUS EXPERIENCE

NAME	CLOCK N°	No 1	No 2	No 3	No 4	No 5	No 6	REMARKS
MIN: ACCEPTABLE PROFICIENCY		B	A	B	B	A	B	
J. Brown	421	E 12 Feb 23 Feb A 23 Feb	C 9 April 27 April A 27 April	D 25 June 9 July B 9 July	E 3 Sept 14 Sept B 14 Sept	A	E 19 Nov 7 Dec	No training needed Job No. 5
R. Green	356	C 26 Feb 2 March B 2 March	C 30 April 18 May A 18 May	E B				
D. White	372	C 5 March 9 March B 9 March	A A	E B	B			
K. Black.	286	E 12 March 23 March B 23 March	E 21 May 18 June C 18 June	C 20 Aug 31 Aug B 31 Aug	C 15 Oct 19 Oct B 19 Oct	D 5 Nov 16 Nov	C 10 Dec 21 Dec	Extra training needed Job No 2.

- PRESENT STANDARD OF PROFICIENCY
- PROGRAMME DATE FOR START OF TRAINING
- PROGRAMME DATE FOR COMPLETION OF TRAINING
- DATE TRAINING WAS ACTUALLY COMPLETED AND STANDARD OF PROFICIENCY ATTAINED

4.26 A training programme

(b) To construct a training timetable for each participant. The programme indicates when each person should start and finish the various parts of his training.

(c) To indicate training progress. When the person can perform the particular job satisfactorily the date is noted. Alternatively, at the end of the training period the standard attained is noted.

All persons do not have the same capacity for learning – some take much longer than others; it must also be accepted that not all training will achieve its objective. But even if only a proportion of the persons trained gain some benefit, the resultant improvement can more than compensate for the effort expended.

4.5. Economic maintenance

Companies spend regularly large sums on maintenance; usually such monies are made available and allocated on a historical basis with adjustments to accommodate price and wage movements. Managements do not always consider the value they are getting for their money or ways in which it could be better spent.

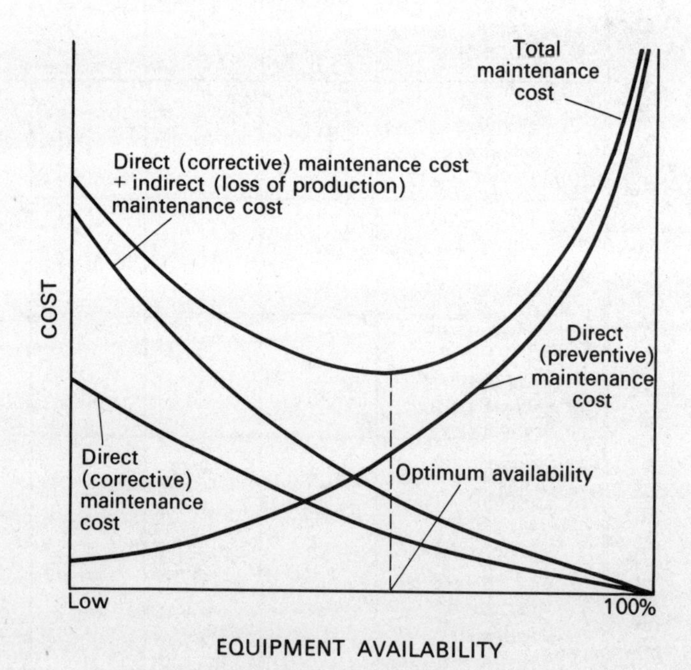

4.27 Mode of determining overall maintenance cost

Maintenance is a service which has a precise economic value to the production process. If this value can be determined and expressed in numerical terms, managements have a means of comparing the cost-effectiveness of different maintenance/production policies. Various maintenance/operational policies were discussed in Section 2.7; if each is to be evaluated with any accuracy realistic cost figures are essential. The direct costs of maintenance must be known, but of equal importance is the indirect cost of lost production, including loss of profit, caused by plant stoppage or non-availability.

'There are various ways in which the direct and indirect maintenance costs can be equated. Figure 4.27 indicates the principle of determining the overall maintenance cost; the figure has been optimized. When the total cost is thus examined, the value of budgetary control of maintenance departments, as normally practised, becomes exceedingly doubtful. It is quite normal for the maintenance manager to forecast his anticipated costs in carrying out maintenance work of all kinds but unless he is also concerned with the consequences of reducing maintenance, he cannot possibly arrive at the optimum maintenance policy'[17].

Although the general principles for the construction of the curves (Fig. 4.27) are appreciated and understood readily, in practice the actual relationship between the various functions may be very difficult to determine. An easier method of assessment is necessary if the maintenance engineer is to compare his efforts with an acceptable standard.

Various authors have endeavoured to formulate a method of evaluating maintenance performance in numerical terms, either as a single overall factor or as a series of factors. Probably the simplest is 'Corder's Maintenance Efficiency Index'[9]; it takes the form

$$\mathrm{E} = \frac{K}{xC + yL + zW}$$

where C = total cost of maintenance, expressed as a percentage of the replacement value of the plant and equipment.

L = downtime due to maintenance causes, expressed as a percentage of the scheduled production hours.

W = waste of materials caused as a result of maintenance responsibility, expressed as a percentage of total output at that stage of the process.

x = total cost of maintenance in the base year.

y = total cost of lost time due to maintenance causes in the base year.

K = a constant such that the value of the expression is 100 for the base year.

E = index of maintenance efficiency. In the base year it will be 100. In subsequent years an index of over 100 indicates an improvement in the efficiency of the maintenance function and values less than 100 represent an unfavourable trend.

Note: The formula assumes that maintenance costs are proportional to plant replacement value.

There are advantages and disadvantages with every method of evaluation. The disadvantage with the above system is that it does not give an absolute value but only a comparative measure of efficiency. Its main advantage is in its simplicity and ease of application.

5
Commercial Systems

5.1. Peg-board charts

Basically, the peg-board chart (Fig. 5.1), is a board drilled with small holes into which pegs can be inserted. The board is drilled with 365 holes from side-to-side, the number from top-to-bottom depending upon the number of machines to be serviced. The horizontal axis represents days in the year, one hole per day, and the actual date represented by each vertical row of holes is marked at the top. The horizontal rows of holes are subdivided into 'weeks' for easy reading, by means of vertical nylon strings. The space between each string is seven holes wide.

The vertical axis represents machine numbers. Every machine is given a reference number, and these numbers are marked on the top of plain pegs, one for each number, inserted in holes on the vertical axis at the left of the chart, but outside the area of the chart itself. Thus, every hole in the board has as its co-ordinates the date and the machine number, for example February 10, and 01/06/19.

Service schedules for each machine are indicated by grooved-top coloured pegs – red for weekly, yellow for monthly, green for quarterly and blue for six-monthly. These are inserted in the board opposite the numbered pegs, under the appropriate date. By examining the pegs that are in the board under any specified date, the operator can see at a glance which machines require which service on that date.

Record cards for each machine are kept in a suspended visible card index system alongside the number pegs. There are four cards per machine – inspection record, inspection frequency, machine record and lubrication record. As each service is carried out and reported, the chart operator removes the appropriate peg, and replaces it with the numbered peg. This peg is attached to a spring loaded reel of thread, so that as the peg is advanced across the board, the thread covers the holes which now contain no coloured pegs. if a specified service is not carried out, the coloured peg is not removed, and acts as a visible reminder that the service remains to be done.

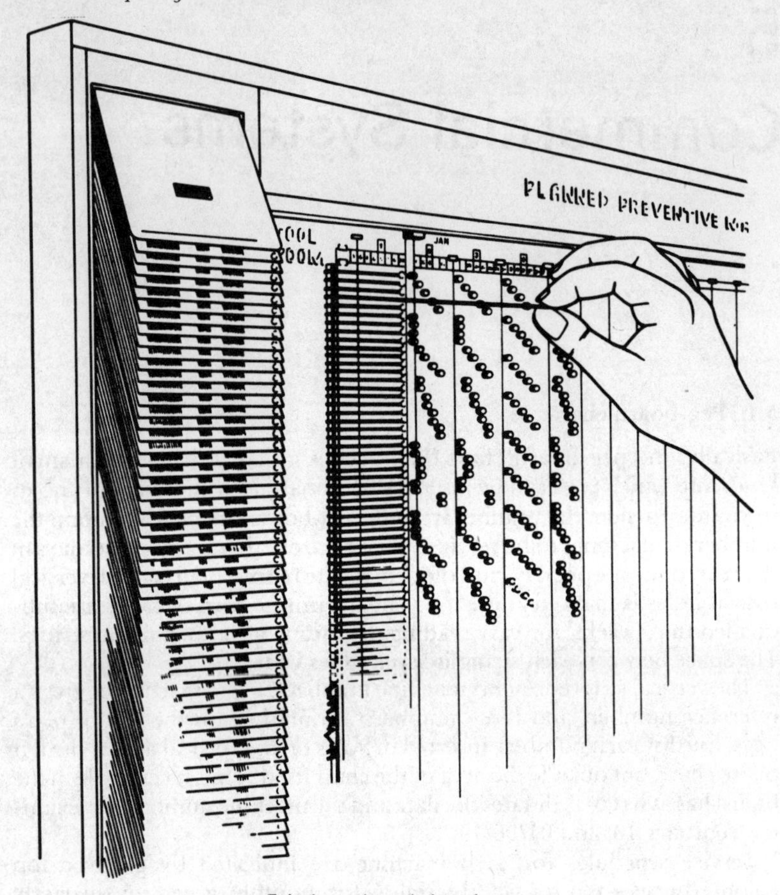

5.1 A peg-board chart in use

The peg-board chart thus provides a visible record of:

(a) what the future programme is,
(b) how well the programme is being maintained, and
(c) what needs to be done, if anything, to get back on schedule.

5.2. Kalamazoo Strip Index

Kalamazoo Strip Index (Fig. 5.2) is ideal for every type of record where the information can be typed or written in a few lines. It is particularly valuable where the information is constantly changing, yet must be kept up-to-date. No matter how often additions or deletions may be necessary, a record on strip index can always be up-to-date, tidy and in the correct sequence. Some typical applications are given below but there are many others each catering for specific user needs:

Address and telephone index.
Index of blue prints, drawings, service manuals.
Index of suppliers and contractors.
Index of stores and spares.
Facility register.
Maintenance programmes.
Maintenance work lists.

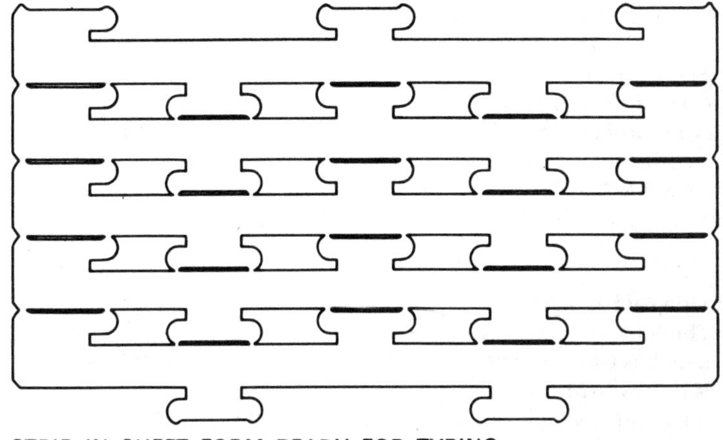

STRIP IN SHEET FORM READY FOR TYPING

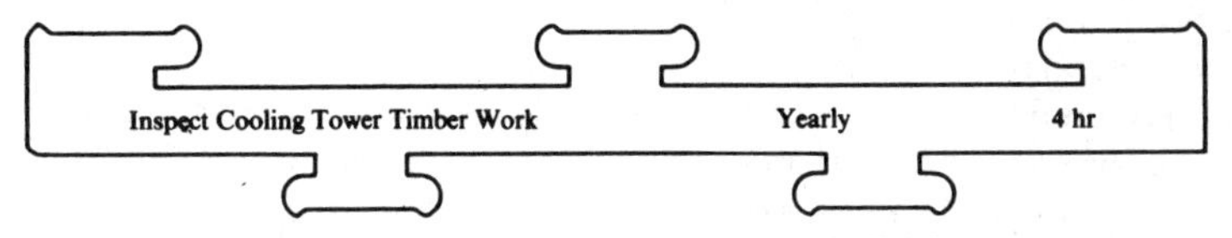

TYPED SEPARATED STRIP READY FOR INSERTION

5.2 Kalamazoo Strip Index system

The information is typed or written on specially shaped strips. These can be printed to any design and are available in several widths, depths and colours. The strips are supplied in sheet form to facilitate typing, when the information has been added they can be cut into individual strips for mounting on to panels. Each panel is fitted with plastic runners which hold securely the strip but allow it to be moved up and down the panel. (See Fig. 2.6.) Strips can be removed or relocated and new ones added, without upsetting the sequence, by sliding the strips adjacent to each other so they touch and a continuous list is formed. (See Fig. 5.3.)

The panels are kept together in loose-leaf binders in the form of books. A binder taking up 76 mm (3 in) of shelf space can hold up to 5000 individual strips.

5.3. Strip Index applied to planned maintenance. Kalamazoo maintenance diary and work lists

The system provides:

(a) a means of showing the details of the planned maintenance programme in a simple week-by-week diary form, and

(b) a quick means of producing the weekly maintenance work lists for distribution.

There are two distinct programming aspects to consider, the long-term and the short-term maintenance work. In general, maintenance work may be broken down into these two categories by considering as long-term those items that are to be serviced on a quarterly, half-yearly or yearly basis. Weekly, fortnightly and monthly items would be classed as short-term.

The long-term diary

1. The loose-leaf binders are split into 52 sections, usually 52 separate panels is most convenient, one for each week of the year. Each section is headed with its maintenance week number.
2. Strips are created for each maintenance job showing the:

   job reference number,
   job title,
   tradesmen to do the job,
   standard time allocated for the job.

   For quarterly jobs, four identical strips would be prepared and loaded at three-monthly intervals in the diary. Half-yearly jobs would appear twice and annual tasks once, in each case being loaded against the appropriate maintenance week number. The weekly work load can be checked quickly by totalling the standard time allocations.
3. Prior to the maintenance week concerned, the appropriate panel is removed from the binder and photocopies made for distribution to the maintenance and production staff (Fig. 5.3). (If required, light coloured panels are available to give the reproduced document a white background. This may be more acceptable for additional notes. Information which appears on the master index strip but is not to be reproduced on the distribution copies, i.e. Standard times allocated, costs, etc. can be masked easily by means of an overlay during the photo-copy process.)
4. The copies (probably several of each panel) are issued by the foreman

to his fitters, electricians, etc. as work lists. The items that each has to service are indicated.

5. The copies of the work lists can act as a job report and be returned with notes and signatures against the jobs actually completed.
6. Since not all the jobs set will always be completed, there must be some method of dealing with the outstanding jobs. There are two courses open:

PREVENTIVE MAINTENANCE Week one January		
Fit Disc to PV 133 Autoclave	Yearly	5 hr
Inspect Cooling Tower	Yearly	4 hr
Samples of Five Mereline Furnaces	6 monthly	5 hr
Insert Gas Plant SO2 Scrubber	4 monthly	2½ hr
Clean Windows in Building 67	3 monthly	8 hr
Inspect N° 5 Set Pot Bottom	3 monthly	1 hr
Inspect Wooden Scrubber Bldg 26	3 monthly	½ hr
Contractors sweep Mereline chimney	3 monthly	
Insert Gas Compressor N° 2	3 monthly	1½ hr
Veck Petrol Meter Pump	3 monthly	2½ hr
Inspect Fans in Bldg 21	2 monthly	1 hr
Stirrer Gland Repacking N° 3 Set Pot	2 monthly	2 hr
Stirrer Gland Repacking N° 4 Set Pot	2 monthly	2 hr
Stirrer Gland Repacking N° 5 Set Pot	2 monthly	2 hr
Stirrer Gland Repacking N° 7 Set Pot	2 monthly	2 hr
Stirrer Gland Repacking N° 12 Set Pot	2 monthly	2 hr
Stirrer Gland Repacking N° 13 Set Pot	2 monthly	2 hr
Stirrer Gland Repacking N° 15 Set Pot	2 monthly	2 hr
Stirrer Gland Repacking N° 19 Set Pot	2 monthly	2 hr
Stirrer Gland Repacking N° 20 Set Pot	2 monthly	2 hr
Inspect PR Ducting Bldg 3	2 monthly	1½ hr
Inspect Swinging Arm of Storage Tank	2 monthly	2 hr
Clean Strainers on Cooling Water Pumps	2 monthly	8 hr

5.3 Copy of a panel ready for distribution

(a) The job can be left until next time it is due for service, providing that the job is treated as a priority, and that there is some means of ensuring that the tradesman is made aware of the fact that it was missed last time.

(b) The job cannot wait until then but must be done as soon as possible; in which case it must be incorporated in work lists to be prepared in the immediate future.

Both of these contingencies can be dealt with easily. In the case of (a), on receipt of the work list indicating that a job has been missed, a red adhesive signal is placed on the next indentical strip in the binder for that particular job. When this is extracted in, say, three months time, a black dot will appear on the resulting photocopied lists. On the return of the list, all missed jobs can be spot-signalled on the master plan using yellow signals which are non-reproducible. (Both the red and the yellow signals can be erased easily without trace from the cards.) At the end of the year, a graphic picture is presented of the success or failure on work load; this can be used to assist programming for the next year.

In the case of (b), it is a simple matter to create a strip for the job and to load it on to a priority list which will be photocopied with the panels of subsequent weeks, the strips being removed and destroyed as the backlog is cleared.

The short-term diary

Less detailed records are required and a very simple solution is thus possible. The binder is divided into only seven sections – weekly, fortnightly phase 1, fortnightly phase 2, monthly phase 1, monthly phase 2, monthly phase 3, monthly phase 4.

One strip is prepared for each maintenance job and loaded as follows:

all weekly jobs in the weekly section;
half the fortnightly jobs in the first fortnightly section;
half the fortnightly jobs in the second fortnightly section;
the monthly jobs spread over the four monthly section.

Each week, the appropriate panels would be extracted and photocopied on a rotation basis.

Week 1. Weekly, fortnightly 1, monthly 1.
Week 2. Weekly, fortnightly 2, monthly 2.
Week 3. Weekly, fortnightly 1, monthly 3.
Week 4. Weekly, fortnightly 2, monthly 4.
Week 5. As Week 1.

By using this simple routine, all short-term jobs are looked at weekly, fortnightly or monthly as required, without the labour and cost of preparing thousands of strips. Yellow signals can be used to indicate missed items as with the long-term system. In carrying out the work, the longest-spaced jobs are completed first.

To summarize

1. The strip index presents an up-to-date appraisal of the planned maintenance work in a clear, simple diary form.
2. Extensive work lists can be prepared without any additional writing or typing.
3. Missed jobs can be catered for simply and effectively.
4. The method has complete flexibility; new items can be added and jobs deleted, maintenance frequency can be adjusted at will. Next year's plan can be created accurately and quickly using the 'bones' from the present year plus experience gained.
5. Control of labour and assessment of realistic work load is simply provided for.

5.4. Kalamazoo Factfinder applied to planned maintenance

A fresh approach has been made to the problem of 'calling up' items of plant included in a plant maintenance system. The whole of a plant maintenance system can be loaded and devised for computer processing. This may well be the ultimate aim of most large concerns, but, in the meantime, a system has been devised which is semi-automatic in nature, which overcomes the main disadvantages of the conventional approach and which will lend itself readily to adaptation for computer operation, if and when this is required.

System description

The system has been devised to eliminate all manual listing and typing in the preparation of maintenance notifications, and to present the plant engineer with a work order which includes:

(a) The name of the plant item to be maintained and its location.
(b) The level of maintenance to be carried out and the chargeable cost code.
(c) A check list itemizing the work to be done.
(d) The estimated time that the job should take.
(e) A list detailing the standard spares which may be needed for the work concerned.
(f) The time and date for which the job is scheduled.

The system also co-ordinates the various trades which may be concerned with work on the same plant item, e.g. fitters, electricians, instrument mechanics, greasers, etc. Minimum staff is required to prepare the work notifications and, as a result, the main effort of the planning office staff can be directed to the analysis of the returns and records, and to the improvement and development of the maintenance coverage. The system depends for its operation upon an edge clipped card specifically designed to allow semi-automatic sorting in a small vibrator box. Figure 5.4a shows a copy of a typical card. The edge pattern on the one side of the card

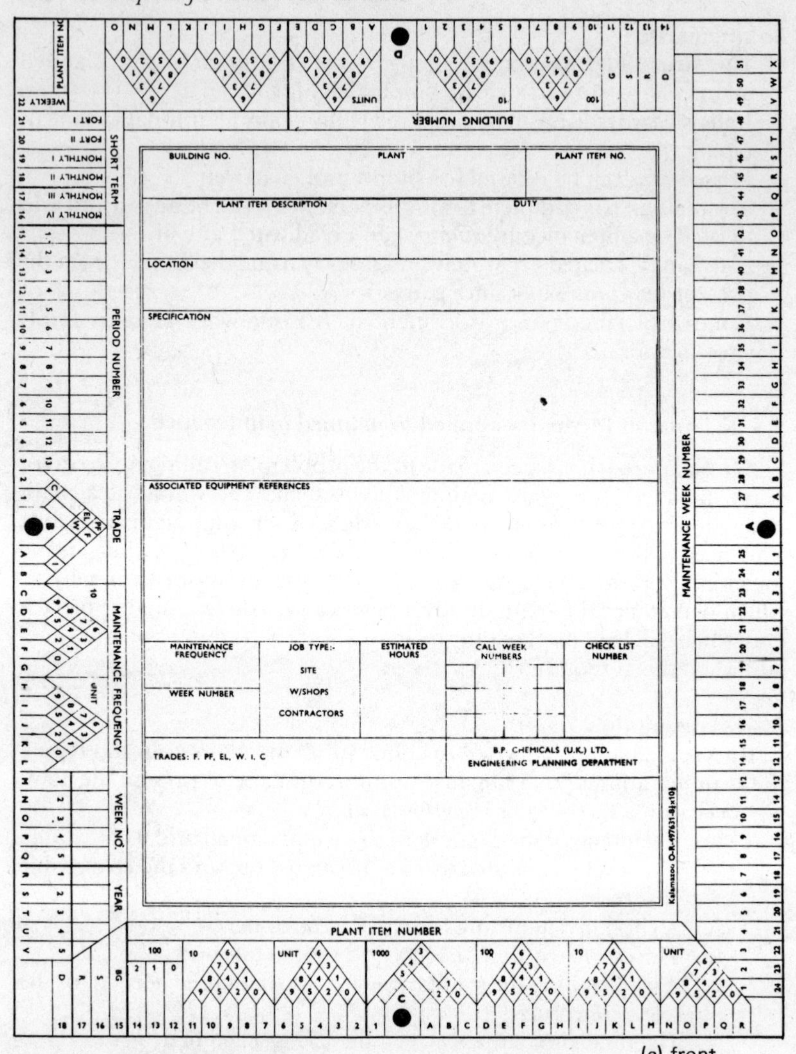

(a) front

5.4 A typical punched edge card as used in the system described

allows notches to be clipped in the edge to indicate the plant item number, building location number, trade involved, week for which the job is scheduled, etc. The centre of the same side of the card allows plant item details to be recorded. A suitable modification to the design could well allow space for the maintenance history to be recorded also. The reverse face of the card (Fig. 5.4b) has only a block of 'boxes' printed down the one side, in which will be typed the indicated information. The

BUILDING NO.: G.110 | PLANT NAME: POLYMERISATION | ITEM NO.: BG/22/ | ITEM NAME: BLOWDOWN | MAINT'CE: Ann B | HOURS: 4½ | SITE | W/S | CON. | DATE AVAILABLE | TIME | PLANT LAB. | W/S LAB.

MAINTENANCE WORK REQUEST FOR at 0800 am/1230 pm

MECHANICAL CHECK LIST

Bldg. No. Item No.
Process
Item
Mntce Cost Code

AUTHORISATION TO START WORK

A permit to work will be needed
type ( ) Sgnd
Job inspected and declared safe
Sgnd
Date Time

NATURE OF WORK

1 Check that the isolation valves are closed, and that all Hamer blinds are in position.
2 Check coupling for wear and renew parts as necessary.
3 Examine all clamps or studs on manhole access, check numbers and condition, with freedom of nuts.
4 Examine valve and pipe mountings and supports for cracks or breakages.
5 Examine oil pipes for leakage.
6 Examine poly main mounting brackets for corrosion, cracks and deterioration.
7 Remove both sight glass mounts and fit replacements.
8 Remove bursting discs, examine vent pipe for blockages, and fit replacement bursting disc.
9 Check gearbox input shaft for end float; report if excessive.
10 Inspect gearbox oil seals for leakage; report if excessive.
11 Remove manifold valves and fit replacements.
12 Inspect bottom bearing (where applicable). Clean and measure wear, if in excess of 1/16" renew.
13 Ensure all guards are effective and secure.
14 Check that plant item is marked with correct item No. Report if otherwise.
15 Despatch removed valves to workshops.
16 Report any defects not specifically covered by the examination.

SPARES LISTED:

COMMENTS: (Continue on back of form)

TIME TAKEN, SKILLED:
TIME TAKEN, UNSKILLED:

DATE: / /19 TIME: SIGN:

(b) Back

main part of this face of the card will contain a form of work order, combined with check list and spares details. This latter information is photocopied on to the card, and the detail heading typed later – as well as the detail along the card edge. Photocopying is a convenient way of producing the cards as many of the plant items will be similar, and thus a standard check list will suffice.

System operation

Once the system is established, a card will exist for each level of inspection required for each item of plant. Each card will be edge clipped in accordance with the week for which the work is programmed. Every week, ten days in advance of the actual work-week for which the listing is required, the cards are run through the vibrator box which has been programmed for that week. The cards thus selected may then be further sub-divided into locations and/or trades, also by means of the vibrator box and the edge clippings.

A list of the work so selected can be made by arranging the cards so that the longer edges overlap sufficiently to 'show' the information typed along the length of the card. Allying this arrangement with a suitable heading, and photocopying on a standard horizontal platen machine, will produce a listing similar to the example shown in Fig. 5.3. This list, being a record of all the maintenance work scheduled for the week required, can then be passed to the plant engineer and process personnel. The plant engineering and process personnel, in the light of manpower and production requirements, can indicate on the list the agreed date and time when the work may be tackled.

The entered up list is then returned to the planning office, where the dates and times are entered on to photocopies of each card.

The now completed work orders can be issued to the plant engineers a few days before the actual work week concerned, and copies sent to all involved parties. In this way, the various trades are co-ordinated to complete their work requirements on a plant item when it is taken out of service, the workshops have prior notification of their loading for the week, and the stores are provided with the opportunity to check the spares required, and, if necessary, supply and deliver items. The items in the check list are marked off as they are attended to, any relevant comments are made on the work order itself, and on completion of the job, the work order sheet is returned to the planning office. All other features of the planning system are operated as usual, the maximum use being made of photocopying.

Conclusion

It is believed that the system described for handling the 'call up' notification is unique, in that once the cards have been established, no listing or writing or searching is involved in the preparation of the actual work orders with the in-built check lists. Furthermore, any week's scheduled work load can be produced at any time without the need to make any adjustments.

In economy of initial setting up cost, and running costs, the system may well rival the ultimate computer applications.

5.5. The Mobil* preventive maintenance system for vehicles and mobile plant

Most large fleet operators run a maintenance scheme. The operator with a few vehicles will normally use the service facilities provided by a local retail service station and will need only a programme chart or diary to show the 'state of maintenance' and to act as a reminder to send vehicles for service. It is probably the operator of the medium sized fleet who will be most interested in the above Mobil system, particularly when he is about to initiate his own maintenance system or to impose a better control in a growing concern.

Equipment survey

Before the system can be installed a complete survey of the equipment in use and the operating conditions under which it operates has to be made. This will enable service intervals to be established for each job, and these are tabulated on a lubrication chart (Fig. 5.5) which can be kept in the office or on the vehicle in a transparent plastic envelope. A carbon copy can be filed for reference in a loose-leaf book containing lubrication recommendations for the whole fleet.

Mobil LUBRICATION PROGRAMME **lubrication chart**

FOR

UNIT	PARTS TO BE LUBRICATED	LUBRICANT RECOMMENDED	REMARKS
VEHICLES			
Albion BRT.548	Engine	Delvac Special	
	Gearbox	Mobilube GX 90	
	Final drive	Mobilube GX 90	
	All grease points	Mobilgrease MP	
			3
		Delvac Special	
		Mobilube GX 90	

5.5 Lubrication chart

Suggested service periods are included on an inspection/maintenance record card (Fig. 5.6). These should be checked against any special requirements of the individual fleet for timing and extra items. Reference to the manufacturer's service manual will give a guide, but any recommendations made in this literature should be regarded as maxima

*The Mobil Oil Co.

CARD No.	**Mobil**	Inspection/Maintenance Record VEHICLE/MOBILE PLANT	FLEET No. REG. No.

Vehicle/Plant	Chas. No.	Make
Date into service	Eng. No.	Engine
Typical Service Periods — Modify as necessary	Trailer No.	Trailer
A. 2 months or 3300 miles	Equipment	
B. 6 months or 10000 miles	Tyre Pressure 1st Axle 2nd 3rd 4th	
C. 12 months or 20000 miles	Tyre Size	
D. 24 months or 50000 miles		

USE	
	IN ENGINE
	IN GEARBOX
	IN AXLE
	IN HYDRAULICS

DATE / MILES

A SERVICE	✓ = OK X = ADJUSTMENT MADE O = REPAIRS NEEDED
1	
2	Change engine oil, check for leaks and auxiliary oil levels.
3	Check hoses and cooling system for leaks, fan belt and anti-freeze.
4	Check battery under load, starter, charging rate and belts.
5	Inspect fuel system and injectors for leaks, check slow running.
6	Check lights, horn, wiper, electrical devices.
7	Check for leaks and top up:— Gearbox and transfer box Torque convertor Axles Steering box Power steering Hydraulic system
8	Lubricate chassis nipples, controls, hinges and locks
9	Fill automatic chassis lubricator.
10	Check clutch adj'm't and fluid level.
11	Check both brakes for adj'm't, fluid level, pressure and leaks.
12	Drain water from air tank and check anti-freeze fluid.
13	Inspect exhaust system.
14	Inspect springs, shackles, dampers tie rods.
15	Check wheel nuts, inspect tyres for condition and hubs for leaks.
16	Check steering for slack, wheel bearings, kingpins.
17	Check instruments, mirrors, door locks.
18	Inspect body work, wings, for damage.
19	Inspect trailer connections.
20	Check operations of auxiliary gear (refrigeration, loading pumps, pipe connections, hydraulic rams, etc.)
21	Check fire extinguishers, safety signs, and tools.
22	Road test for noise, steering, brakes, smoke on full throttle acceleration.
23	Set idle speed if necessary.
24	
25	
26	
27	
28	

Form 1980(10/65)

5.6a Inspection/maintenance record card (front)

CARD No.	**Mobil** Inspection/Maintenance Record **VEHICLE/MOBILE PLANT**	MAKE	FLEET No.

	B SERVICE COMPLETE A SERVICE FIRST	DATE / MILES				
1						
2	Change engine oil filters.					
3	Service plug points, distributor and timing. or injectors (test).					
4	Check cyl. head and manifold nuts.					
5	Adjust valves and check for wear in rocker gear.					
6	Service air cleaners and breathers.					
7	Clean fuel filters.					
8	Check oil pressure hot and cold.					
9	Change axle, and gearbox oils.					
10	Check brake lining thickness. ¼ ½					
11	Check brake pipes and hoses.					
12	Tighten U bolts, and axle shaft nuts.					
13	Check universal joint flange bolts and engine mountings.					
14	Check chassis rivets and body bolts.					
15	Check steering alignment.					
16	Check headlamp alignment.					
17	Test foot brake efficiency and report.					
	Test hand brake efficiency and report.					
18	Road test and report.					
19						
20						
21						
22						
23						

Special lubes or fluids required. - - - - - - - - - - - - - - - -

✓ = OK X = ADJUSTMENT MADE O = REPAIRS NEEDED

	C SERVICE COMPLETE A AND B SERVICES FIRST	DATE / MILES			
1					
2	Check water temperature and if nec'y replace thermostat.				
3	Check charging voltage and regulator.				
4	Clean dynamo and starter brush gear.				
5	Check injector timing.				
6	Check valve timing.				
7	Check compressions with gauge and report. 1 2 3 4 / 5 6 7 8				
8					
9					
10					
11					
12					

	D SERVICE COMPLETE A, B & C SERVICES FIRST	DATE	MILES
1			
2	Remove diesel injector pump and test.		
3	Strip, inspect and repack wheel brgs.		
4	Reverse flush radiator and cylinder block.		
5			
6			

5.6b Inspection/maintenance record card (back)

and used under easy operating conditions. For example, high-mileage vehicles may be able to operate with the maximum mileage intervals quoted, but for low-mileage vehicles shorter intervals will be necessary or it may be better to use a time period. For equipment not fitted with mileage recorders or on pumping vehicles, the intervals should be in terms of hours. Where the engines are run at low temperatures, intermittently, or in dusty conditions much shorter periods should be established.

Inspection/maintenance record

A separate inspection/maintenance record card (Fig. 5.6) is made out for each vehicle or item of plant. This becomes an instruction to the mechanic, detailing the items he should inspect, adjust or lubricate. After the inspection the entries on the card show what other jobs are necessary. It will also show whether the same fault occurs repeatedly so that appropriate action can be taken.

The operator will already keep a record of fuel and engine oil issued either per day or per vehicle/engine. This record will show the mileage or hour-meter reading at the end of each day or shift. In addition, the driver (of a goods carrying vehicle) is required by law to keep a record of his hours of work. Suitable forms can be obtained from stationers and from some vehicle operators trade associations. In some cases, the reverse of these sheets can be used by the driver to report any faults requiring attention.

Workshop attention

When the inspection/maintenance record card or the driver's report indicate that workshop attention is necessary, details are filled in on a job card (Fig. 5.7). It is not necessary to use this job card for simple jobs; it should be used only for larger jobs requiring workshop attention. The job card is not only an instruction to the workshop but also a record of the parts used and their cost, and the cost of labour with the names of the mechanics carrying out the work. Hence each job can be costed individually and the cost allocated to the appropriate budget heading.

Detailed periodical vehicle/plant costing

From the maintenance records, the records of fuel and oil issues, the mileage or hours run, and the details on the job cards, a complete cost analysis can be carried out to show the cost history of each vehicle or item of plant. A suggested lay-out for this analysis is shown in (Fig. 5.8).

This analysis is so arranged that comparisons can be made from year to year, between different makes and types of vehicle or plant, between different drivers or different depots or areas. Such information will help to decide future action when considering replacements of plant or servicing periods.

Mobil **job card**

JOB NO.

PART NO.	MATERIALS	QUANTITY	COST

LABOUR ____ HRS. @
__________ HRS. @
OVERHEADS

TOTAL COST

COST ALLOCATED

(a) back

Mobil job card

JOB NO. ______
REG NO. ______
FLEET NO. ______
DEPOT ______
FITTER ______

CHASSIS NO. ______
ENGINE NO. ______
MILEAGE
HOURMETER } RDG. ______
DATE ______

INSP. OR REPORT DATE	ITEM NO.		INITIAL AFTER COMPLETION

DATE	FITTER	HOURS	FITTER	HOURS

TESTERS REPORT

TESTERS SIGNATURE DATE

5.7 Job card (b) front

VEHICLE COSTING SHEET

VEHICLE COSTING SHEET

MAKE AND TYPE ______________________

LOAD CAPACITY ______________________

COST ______________________

ESTIMATED LIFE ______________________

Weekly costs \ Week ending									
Miles/Hours									
Fuel									
Lubricants									
Maintenance									
Repairs									
Spare Parts									
Contract Repairs									
Weekly Running Cost									
Wages									
Proportion of									
Insurance									
Licence									
Weekly Standing Costs									
Total Costs									
Revenue									

5.8 A vehicle costing sheet

5.6. The Mobil basic system*

This is a simple two card system which can be operated in different ways to suit local conditions. The first step is to transfer from the lubrication survey to a machine register card (Fig. 5.9) all the items which need lubrication attention. A separate card is made out for each machine, or section of machine in the case of large pieces of equipment such as production transfer machines. Each card identifies the machine by name,

*All Mobil cards shown are specimens only and are not freely available from Mobil. They are subject to alteration at any time.

number and location. On the reverse of the card details of repairs, overhauls and modifications are entered to provide a complete history of the machine.

All lubrication tasks are transferred from the lubrication survey. First the daily tasks to be undertaken by the operator, then the weekly, monthly, etc. tasks which are the oiler's work and finally any lubrication tasks such as re-packing electric motor bearings which are the responsibility of the electrical and other maintenance functions. These three groups of tasks are clearly delineated. If there are sufficient entries to fill more than one card then each type of task can be entered on a separate card. In this case each card must be clearly marked – operator's tasks – oiler's tasks – electricians' tasks.

When completed, the machine register card is put inside a transparent plastic envelope or heat sealed between plastic sheets and hung up on or near the machine. A readily visible indication of all lubrication tasks is then present at all times.

It might be thought necessary to take a carbon copy of the machine register cards, either to guard against loss or to have a master set of all machine register cards in the maintenance foreman's office or oil store. However, in this case, the lubrication survey is the real master copy and all cards are derived from it.

The second card is the lubrication record card (Fig. 5.10) and this too is identified with the machine name, number and location. Across the top it carries an alphabetical index which corresponds to the vertical index on the machine register card. This card is also put into a plastic pocket and kept on or near the machine.

How the Mobil basic system works

This system does not incorporate any form of oilers job card or route card and is operated on the basis that the oiler has specific routes each day which ensure that he attends to each machine once a week. The daily tasks are the operator's responsibility and as in a machine shop continuing production can result only from properly maintained machines, production incentives invariably ensures that daily and shift lubrication functions are properly attended to.

Each day, after preparing his lubrication trolley, the oiler proceeds on his established route for that day stopping at each machine in turn. From the machine register card he determines which tasks to perform and having done so enters the lubrication record card accordingly. As his route is designed to cover each machine once a week he will always perform the weekly tasks. In addition, on every fourth week, he performs the monthly tasks as well. The specimen lubrication record card shows this.

When he drains and refills a reservoir or gearbox he will enter the amount of oil used on the lubrication record card. On the basis of a weekly minimum period, the lubrication record card will give a complete history of lubrication for each machine for a three year period and this, together with relevant information on the reverse of the machine register card, enables annual statements to be prepared for each machine detailing lubrication costs and recommendations for more efficient operation.

Equipment

To install the Mobil basic system of machine lubrication control, all that is needed is a supply of machine register and lubrication record cards and the plastic envelopes to house them and attach to each item of plant. In those cases where it is desired to have a duplicate set of machine register cards, they can be housed in a simple filing box.

Limitations of the system

This simple system has limitation. It does not identify easily situations needing attention due to excessive or otherwise non-standard lubrication requirements. Although a leaky hydraulic system may be the cause for the weekly attention 'tick' being replaced by a figure to indicate excessive rate of top up – there is no means of throwing this situation up for attention by maintenance men. The only way in which excessive or scanty consumptions or other unusual circumstances can be registered is by the use of a squawk sheet as explained later in the Mobil System A (page 170).

MACHINE: Feedick Radial Drill
NUMBER: 0102

LOCATION: No.5 Machine Shop

Mobil LUBRICATION PROGRAMME **machine register**

	ITEM	NO. OF POINTS	LUBRICANT	APPLICATION METHOD	APPLICATION FREQUENCY	CHANGE/ REPACKING FREQUENCY	CAPACITY IN GALS /LBS	
A	Hydraulic system	1	DTE Light	Can	W			
B	Head Gearbox	1	Vactra BB	"	W			
C	Main Gearbox	1	Vactra BB	"	W			
D	Elevating Gear	1	Compound EE	"	W			
E	Elevating Screw	1	Compound EE	"	M			
F	Spindle Bearing	2	Mobilplex 47	Gun	3M			
G	Head Gearbox	1	Vactra BB	Can		6M	10	
H	Main Gearbox	1	Vactra BB	"		6M	10	
J	Elevating Gear	1	Compound EE	"		6M	5	
K	Hydraulic system	1	DTE Light	"		12M	40	
L								
M								

Code No. 8022

5.9 Machine register card used in the Mobil basic system

It is also difficult to incorporate a system of consumption records either by machine or by grade of lubricant as the record of consumption is spread out all over the machine shop.

These and other shortcomings are known and are overcome by the following Mobil systems.

MACHINE: ____________ LOCATION: ____________
NUMBER: ____________ ____________

Mobil LUBRICATION PROGRAMME **lubrication record**

✓ Indicates item has been lubricated X Indicates lubricant has been changed Figure indicates quantity added to Reservoir, in gallons

DATE	INITIALS	A	B	C	D	E	F	G	H	J	K	L	M	DATE	INITIALS	A	B	C	D	E	F	G	H	J	K	L	M

Code No. 8023

5.10 Lubrication record card used in conjunction with the machine register card in the Mobil basic system of machine lubrication control

5.7. The Mobil system A

From the lubrication survey, two machine register cards are typed out for each machine (or section of machine). The top copy becomes the oiler's working reference and the carbon copy is fixed in a plastic cover to the machine in a prominent position. In some cases, there may be more lubrication tasks involved than there are spaces on the card. This can easily be overcome by making out additional supplementary machine register cards.

Each card (Fig. 5.11) identifies the machine and its location. It details the items on the machine needing lubrication, the number of lubrication points, the lubricant to be used, method of application, the service period, estimated extended monthly usage of those lubricants for which it is desired to keep records, the repacking/change period for grease packed bearings/circulation systems and gearboxes, and the capacity of these units.

MACHINE: Cincinnati Vert Broach
NUMBER: 8501
LOCATION: No. 5 Shop

Mobil Machine Register

Ronco Organisation

ESTIMATED MONTHLY CONSUMPTION-GALLS.			
D.T.E. Lt.	1	Vactra No. 4	1

ITEM	PART TO BE LUBRICATED	PTS	LUBRICANT	METHOD	SERVICE D	SERVICE W	SERVICE M	CONS D	CONS W	CONS M	CHANGE PERIOD	CAP
	Pilot pump - piston rods Hyd. oil pump	4	Mobilux 2	Gun	X							
1	Table Hyd. Pistons	1	D.T.E.Med.	Can		M						
2	Ram and Table Ways	1	VactraNo. 4	Can		M			½	1	12M	10
3	Hydraulic System	1	D.T.E.Lt.	Can			1M			1	12M	150

5.11 Mobil machine register card

This information is taken from the lubrication survey. However, only that which is the responsibility of the oiler and the machine operator is transferred. Details on motor bearings are therefore in this instance omitted. They will be transferred to a separate electrical maintenance schedule.

The first items to be transferred to the machine register card will be the items requiring daily attention – the responsibility of the machine operator. Figure 5.11, shows this. These are marked 'x' under the service period heading D – to indicate they are in fact daily tasks.

Then the tasks requiring weekly attention are transferred followed by those needing monthly attention. These are given an item serial number and are of course the oiler's responsibility. The weekly tasks are given a letter under the service period heading W to indicate the day of the week on which they should be completed (e.g. M = Monday). It is preferable that all weekly tasks on each machine should be scheduled for the same day. Similarly, the monthly tasks are given an entry under the service period heading M. In this case – 1st M – indicates that the task should be completed on the first Monday in the month. If calendar months are used, the fact that there are four and five week months has no practical significance. It is also preferable that the monthly tasks on each machine should be scheduled to coincide with the weekly tasks. Then both the weekly and monthly tasks can be completed together with obvious benefits in terms of time spent on the job.

5.12 Roneodex desk wallet

The machine operator knowing he is responsible for all daily lubrication tasks has a ready reference on his machine and if these are completed at a set time each day, it is easy to set a regular working pattern to ensure regular attention to them.

The desk wallet

The oiler's reference copies of the machine register cards are fitted into five desk wallets, Fig. 5.12, the size of which will depend upon the number of machines and the type of plant involved. These are clearly labelled – Monday to Friday. Each desk wallet is the oiler's job card and route card for that day. The machine register cards can be protected by PVC sheets.

It is the actual make up of the desk wallets which is the crucial part of the system. The machine register cards must be inserted and the lubrication tasks allocated to ensure the most economical use of oiler time. In some cases, there might be a requirement for one type of lubricant on a few machines only. The obvious solution is to make up the desk wallets so that this requirement is discharged completely on one day to preclude the necessity to draw, carry and dispense a 'speciality' more frequently than necessary.

In this example lubrication periods are limited to daily, weekly and monthly, and although there is a requirement on the machine register card that two circulation systems should be drained, flushed and refilled yearly, there is no indication of when this must be done. Tasks such as these are normally carried out when the machine is stationary either at weekends or other holiday periods. They are entered in special weekend desk wallets or integrated in the mechanical maintenance workload schedule.

In the same way, three monthly, six monthly or other 'odd' periodicity tasks can be accommodated in special desk wallets.

The lubrication record card

Each machine register card having operator's and oiler's tasks on it – also has a mating lubrication record card. This card is identified with the machine by name and number (Fig. 5.13). The task numbers from the machine register card are transferred across the top of the lubrication

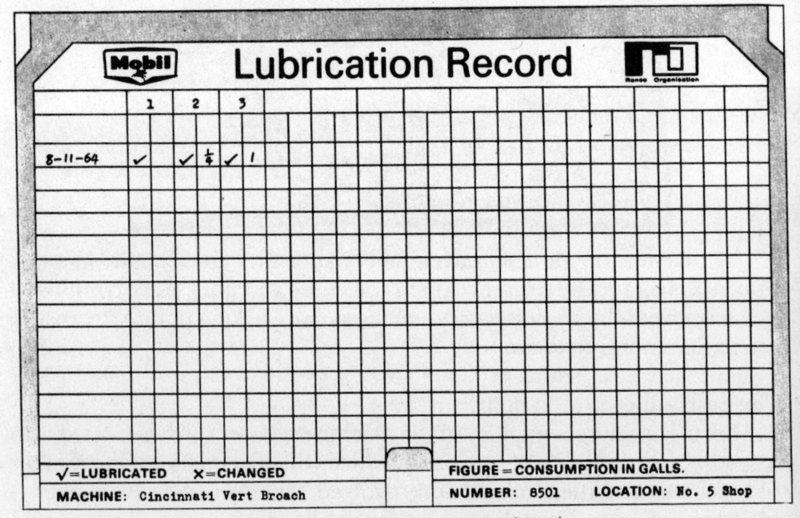

Mobil

Lubrication Record

Renee Organisation

	1		2		3	
8-11-64	✓		✓	¼	✓	1

✓=LUBRICATED X=CHANGED FIGURE = CONSUMPTION IN GALLS.

MACHINE: Cincinnati Vert Broach NUMBER: 8501 LOCATION: No. 5 Shop

5.13 Mobil lubrication record card

record card. Each lubrication task is therefore identified by machine and item number.

The lubrication record cards are also fitted into the desk wallets so that each machine register card lies flat above its mating lubrication record card.

How the Mobil system A works

Each day the oiler draws from the maintenance department clerk the desk wallet relevant to that day. Then after preparing his lubrication trolley he performs all lubrication tasks in the order indicated. (The machine register cards could be loaded by indicating the task day with a tick or cross and then ensuring that these cards were included in the correct day's desk wallet – but this is not recommended since indicating the actual day is of great assistance in loading the desk wallets.)

After completing each task the oiler enters the date and ticks the relevant square on the lubrication record card and in those cases where it is necessary to record the amount of oil used, this is entered in the second square as indicated. He will know which consumptions to record since these will be indicated on the top of the machine register card.

Although the oiler is not expected to perform maintenance tasks his job does include reporting unusual circumstances, such as hot bearings, leakage from hydraulic lines, etc. He carries a squawk sheet pad, and when he sees or has reported to him items needing mechanical or electrical attention, he fills in a sheet from the pad giving brief details of the circumstances and slips it into the pocket holding the lubrication record card. The design and size of the squawk sheet is entirely dependent upon local circumstance and requirements. Figure 5.14 shows a possible design – the simpler the better.

After completing all lubricating tasks, the oiler returns the desk wallet to the maintenance department clerk, who checks the amount of oil used against the estimated quantities entered on the machine register card. If there are significant variations then he will make out a squawk sheet, slip it into the lubrication record card pocket and insert a red signal in the base of the card. He also inserts a red signal in the base of cards already having a squawk sheet in them.

The desk wallet is then passed to the maintenance foreman. The red signals indicate where action is necessary. The squawk sheets are extracted and the desk wallet returned to the maintenance clerk who extracts the red signals. The faults will then be rectified as a mechanical or electrical maintenance function, or it may be that there is need to adjust the estimated consumption figure on the machine register card.

Stock control

The Mobil system not only ensures that all lubrication tasks are completely regularly but also gives a simple stock control procedure.

Squawk Sheet

MACHINE: Cincinnati Vert Broach
NUMBER: 8501
LOCATION: No.5. Shop

HYD. Pump Leaking

DATE: 8-11-64 OILER: A.B.Smith

5.14 Suggested design for a Squawk sheet. A pad of these sheets is carried around by the oiler. When he notices any unusual circumstances he fills in one and inserts it in the relevant lubrication record card pocket in the desk wallet

The oiler draws oils and greases from the oil store against a normal requisition issued by the maintenance clerk or maintenance foreman. Each day the oil storeman transfers information from these requisitions to a set of 'lubricant stock sheets' in a simple loose-leaf binder. The design and size of the lubricant stock sheet is a matter for individual consideration. Figure 5.15, shows how a lubricant stock sheet might be laid out. There is a sheet for each grade of lubricant stocked. On these sheets the

Lubricant Stock Sheet — May65

LUBRICANT: D.T.E. Light   EST. MONTHLY CONS: 80

DATE	REQ. NUMBER	ISSUES	ACC. ISSUES	RECEIPTS	ADV. NOTE
B/FWD:				85	
1	457	5	5		
1	462	5	10		
3	473	10	20		
				45	12684
			85	130	
			BALANCE C/FWD:		45

5.15 Suggested layout for lubrication stock sheet. One sheet is used for each grade of oil and grease kept and used in the store. In many plants, existing stock control sheets or cards can easily be adapted to serve this purpose

storeman enters the requisition number, amount drawn and the accumulated quantity drawn that month. A separate sheet is made out for each month. At the top of the sheet the estimated monthly consumption is entered being the total from all the machine register card entries. For greases and other products, the quantities of which are not recorded on the machine register cards, the estimated monthly consumptions can be established by experience over a few months.

At the end of each month the oil storeman makes up each lubricant stock sheet, enters the balance of product which should always tally with actual stock and passes the sheet to the maintenance department clerk, who compares actual usage with estimated consumption and in the event of serious discrepancies, compares the sheet with lubrication record cards for that month. If the lubrication record cards have not already indicated a reason for the discrepancy, the reason must be sought elsewhere. This becomes a maintenance department task.

After satisfying himself that the lubricant stock sheet is either in order (or explained) the maintenance clerk orders more lubricant, signs or stamps the sheet and returns it to the oil storeman for filing at the back of the loose-leaf book in which the sheets are kept. He will then be able to maintain annual consumption figures as required.

5.8. The Mobil system B

From the lubrication survey two machine register cards are prepared for each unit or part of unit requiring periodic lubrication attention. The machine register card lists the items on each unit needing attention by the

MACHINE: Sykes Gear Shaper
NUMBER: 2345
LOCATION: M/C Shop Gear Sect.

Mobil Machine Register

SERVICE PERIOD	PART TO BE LUBRICATED	NO. OF POINTS	LUBRICANT	METHOD	NOTES
D	Recip Drive Brg.	1	Vactra Hy Med.	Can	
	Recip Sleeve Brg.	2	"	Can	
	General Lubrication	6	"	Gun	
	Slides	1	Vactra No. 4	Can	
W	Main drive g'box	1	Vactra Hy. Med.	Can	Drain, clean and refill every 2 years
	Cutter feed worm gears	2	Vactra No. 4	"	
	Intermediate g'box	1	"	"	
	Tool head g'box	1	"	"	
	Work head intermediate g'box	1	"	"	
	Work head rotation gears	1	"	"	
	Bottom bevel gears	1	"	"	

5.16 Mobil machine register card

oiler and machine operator. It is important that all daily, weekly and monthly tasks are grouped. There is no reason why three-monthly, six-monthly and yearly tasks should not also be entered on the card. However, the example in Fig. 5.16 shows daily and weekly tasks only. The only items not transferred from the lubrication survey are those which are the responsibility of the electrical or mechanical maintenance section. These will be transferred to an electrical or mechanical maintenance schedule.

One copy of the machine register card is housed in a plastic envelope and either hung upon or near the machine. Alternatively, it can be heat sealed in acetate covers and pinned to the machine. The machine operator is responsible for the daily tasks and this card serves as a ready reference to the items needing attention. There is no record of these daily tasks having been completed, but signing a card or any similar type record is equally no proof that the job has been done. In production machine shops, and it is in this type of plant that the daily tasks are usually made the responsibility of the operator, the incentive of continuous production for bonus purposes is normally sufficient to ensure that the daily tasks are effectively carried out.

The card holder units

The carbon copy of the machine register card and a mating lubrication record card (Fig. 5.17), are mounted in an individual pocket type card holder unit (Fig. 2.7) in such a way that the paired cards can lie open and flat simultaneously. The card holder unit takes 61 pairs of cards and is housed in a unit container. Record cabinets are available to house 6, 12,

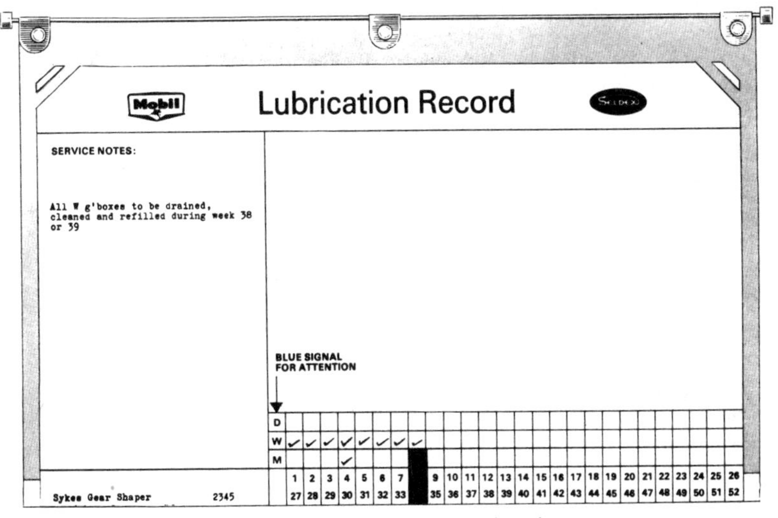

5.17 Mobil lubrication record card

15 or 18 card holder units, each one holding 61 sets of cards (Fig. 2.7). The number of unit containers or size of record cabinet will therefore be dependent upon the number of units requiring lubrication divided by 61. Whether unit containers or record cabinets are used will depend upon the way in which the Mobil system is employed.

Lubrication record card

The lubrication record card is designed to last for one year (52 weeks), at the end of which it can be scrapped or retained for background statistical information. Besides carrying on its visible edge the description and abbreviated detail of the machine to which it refers, the lubrication record card carries a weekly scale and a visible red signal which is moveable along the weekly scale. Above the weekly scale there is a panel denoting the daily, weekly and monthly tasks – in this case by D, W and M. As the daily tasks are carried out by the machine operator the D line is in-operative. Lines for 3M and 6M, etc. can be included if these frequencies are included on the machine register card.

It is not essential that the year covered by the card should be a calendar year – the 52 week period can start any time.

The remaining portion of the lubrication record card is used for supplementary information which is carried forward as necessary from the previous year's card. Such things as attention at 2 yearly periods to motor bearings – details of sample reports on circulation oils, an indication of consumption of lubricants, etc. Much will depend upon the type of information it is required to keep. This serves as the basis for an annual report by the maintenance/lubrication staff to management. Other items which might be included are consumption records, details of change of grade of lubricant, or period of application, details of application, method modification, etc.

The oiler's job cards

The oiler's job card is the trump of the whole system. As the oiler works a five day week there are 260 cards in successive order marked up Monday to Friday through the weeks from No. 1 to No. 52. These cards are loaded by marking on them the machine number and the tasks to be done by the oiler – W for the weekly, M for the monthly and so on (Fig. 5.18a). The job cards are loaded so that each machine has all its necessary lubrication functions carried out within the allotted periods.

These job cards are kept in numerical order in a simple drawer unit. Each card is either sealed between plastic sheets or kept in a tightly fitting transparent envelope to keep it clean. Against each machine number there is also a number of columns, which are used to indicate the quantities of oil used to complete the tasks.

The reverse of the job card (Fig. 5.18b) is used to indicate the

(a)

Monday—Week 4

MACHINE	PERIOD		GALLONS				
			D.T.E. Lt.	Vactra Nº 4	Vactra Hy. Med.	Vacuoline 1405	
346	W	M	1				
857	W			½			
964	W	M	1		½		
1078	W					¼	

(b)

Monday—Week 4

Machine: Norton Grinder Number 765

Drain, clean and refill hydraulic system.
15 gallons Mobil D.T.E. Oil Light

REPORT:

857 hyd.
System
leaking.

5.18 (a) Oiler's job card. On this he indicates in chinagraph or other removeable crayon that he has done the tasks allotted.
(b) Reverse of the oiler's job card. This side indicates non-routine tasks for the day and also serves as a squawk sheet

non-routine jobs which must be done on that day such as changing oil in reservoirs and gearboxes, repacking bearings, etc.

How the Mobil system B works

As previously, this description is based on simple premisses – one oiler and a five day week. Daily tasks are performed by the machine operators. It shows one simple way of operating the system. Others are, of course, possible and practical.

Every morning after replenishing grease guns and oil containers on his lubrication trolley, the oiler collects from the maintenance department clerk his lubrication job card for the day. This is made out in such a way that the oiler is given the most economical route for his day's work. The job card also serves as a route card.

As he progresses from machine to machine, the oiler can see from the job card which jobs are to be done. For example, if he has a W against a machine number then all the weekly jobs must be done on that machine. The detail of these he sees on the machine register card which is on the machine. Having done the jobs he marks off the W on the job card with a chinagraph pencil or similar marker. At the same time, he marks in the final columns the amount of oil he has used to do the job. These columns will be headed to suit the grades of oil for which it is required to keep consumption records. The oiler will also indicate on the reverse of the card any unusual circumstances which he feels merit attention such as leaking joints, hot bearings, etc.

On completing the day's work, the oiler returns his trolley to the oil store and his job card to the maintenance department clerk. The clerk will then (during the following day) refer to each relevant lubrication record card in the card holder unit. First he will move the red signal to the number corresponding to the current week and mark off the completed tasks with a tick. As these signals are visible it is easy for the maintenance foreman to see at a glance which lubrication tasks are more than a week overdue. The clerk will also note whether the volumes indicated on the job card tally with those indicated on the lubrication record card. In this case, the lubrication record card will carry an indication of the amounts of oils which should be used to perform the indicated tasks. If the amounts recorded on the job card differ materially from the guide lines marked up on the lubrication record card then obviously something is wrong (joint leakage, etc.) and the clerk will insert a blue signal in a predetermined position. He will also insert the blue signal if the oiler has indicated other troubles, i.e. hot bearings, in writing on the reverse of the job card. In either event, in addition to inserting the blue signal, the clerk will make a note describing the trouble and tuck it in the card. Then when the maintenance foreman makes his daily check of lubrication troubles all he has to do is cast his eye down the visible signals. Blue ones indicate trouble. He opens the card – extracts the note, removes the blue signal, and assigns the job to a maintenance/repair function. At the same time, he can immediately see which machines are not lubricated to schedule and take the necessary remedial action.

After transferring information from the job cards to the lubrication record cards, the clerk cleans off the chinagraph pencil markings and returns the card to the back of the drawer unit where that particular card will come up again for action one year later.

The system also permits cumulative consumptions for specific circulation systems or reservoirs on specified machines to be recorded. All that is involved is for the clerk to add consumptions from the job card to a cumulative total on the lubrication record card. In this way, it is very simple to keep annual consumption totals split down by machine.

As it becomes necessary to drain and refill a system or perform some task other than the routine inspection or to top-up indicated on the machine register card, this is written into the reverse of the job card as a specific task. When completed a note to this effect should be added to the lubrication record card by the clerk.

6

Projects

6.1. Resource planning project

A process consisting of five main interconnected units is to be shut down for maintenance, repair and modification. The personnel to carry out this work are to be drawn from the central maintenance pool but only the following persons are available for this particular overhaul:

1 mechanical fitter	1 welder
1 electrician	1 lagger (thermal insulator)
1 instrument fitter	1 chemical cleaning/degreasing operator

The work sequence on each unit is set out below together with the estimated times (in days) of the individual work elements.

Work Sequence	Unit				
	A	B	C	D	E
1	Instrument Fitter (2)	Mechanical Fitter (2)	Mechanical Fitter (2)	Mechanical Fitter (3)	Electrician (1)
2	Welder (2)	Instrument Fitter (6)	Electrician (4)	Electrician (3)	Mechanical Fitter (1)
3	Electrician (4)	Mechanical Fitter (4)	Instrument Fitter (5)	Lagger (3)	Lagger (3)
4	Welder (3)	Welder (4)	Mechanical Fitter (3)	Chemical Cleaner (2)	
5	Lagger (2)				

Estimated work times shown in brackets.

On completion of the work, a further period of one day is required for the simultaneous recommissioning of all the units. Assuming:

(a) that the tabulated work sequences for each of the individual units is adhered to, and

(b) that the individual tasks (work elements) once started will be continued – they will not be split,

determine:

1. The minimum time in which the complete overhaul, including recommissioning, can be carried out.
2. The work programme applicable to each unit.
3. The work programme for each of the personnel involved.

Suggested procedure

The situation described in the problem is typical of many continuous operating plants that are scheduled for, perhaps, three weeks shut down each year for overhaul. It is vital that the period of downtime is not exceeded, indeed, if it is possible to complete the work and start up the plant again in less time, considerable financial benefits often result.

Although this exercise deals specifically with the allocation of personnel, the same principles and procedure could be applied equally to the scheduling of other resources, i.e. the allocation of plant and equipment, compressors, cranes, generators, etc. that are in common use among various sites. Probably the easiest, quickest and best understood method of programming is by means of a Gantt chart, and is the one that forms the basis of the following procedure:

1. Calculate the total work time required to overhaul each unit – the sum of the respective work elements.

   Unit A: 2 + 2 + 4 + 3 + 2 = 13 working days.
   Unit B: 2 + 6 + 4 + 4 = 16 working days.
   Unit C: 2 + 4 + 5 + 3 = 14 working days.
   Unit D: 3 + 3 + 3 + 2 = 11 working days.
   Unit E: 1 + 1 + 3 = 5 working days.

2. Calculate the work allocation for each tradesman.

   Mechanical fitter 15 working days
   Electrician 12 working days
   Instrument fitter 13 working days
   Welder 9 working days
   Lagger (thermal insulator) 8 working days
   Chemical cleaner/degreaser 2 working days

3. Examine both the above lists to determine the largest element of time.

In this instance, it is the overhaul time required for the Unit B, i.e. 16 days. Thus the absolute minimum time to complete the programme cannot be less than 16 days + 1 day for recommissioning = 17 days total.

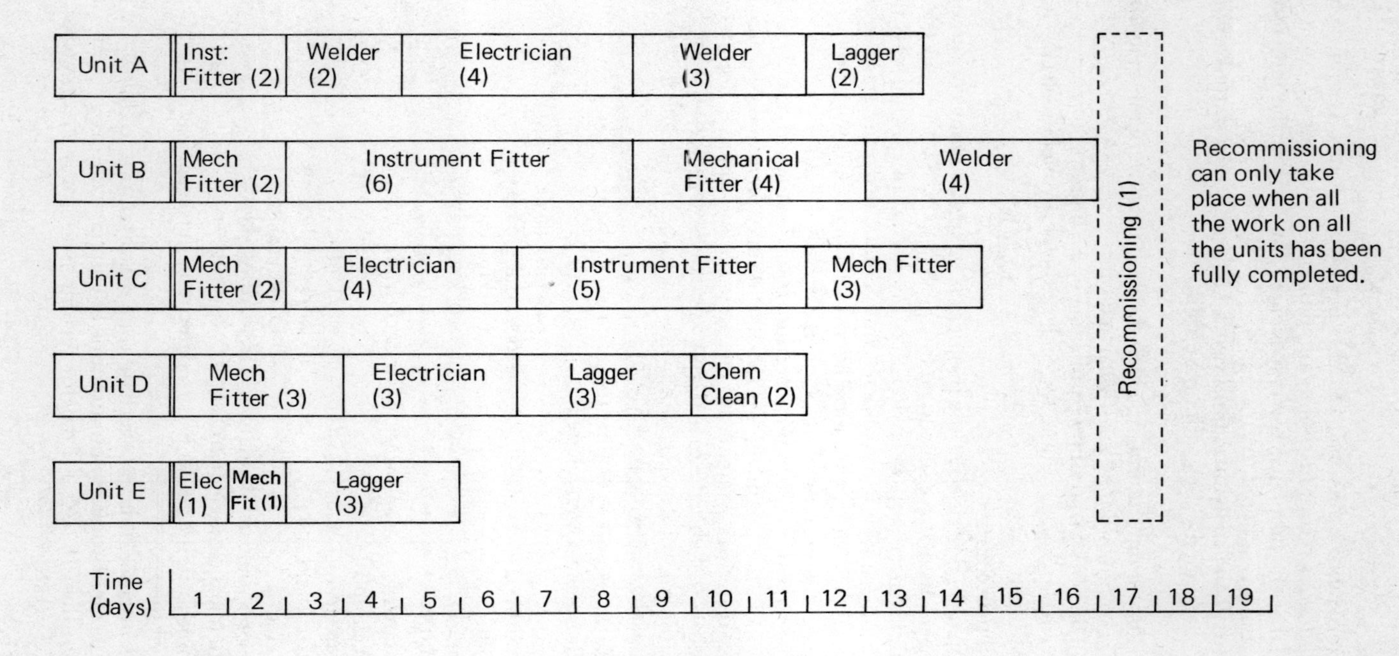

6.1 (a) The work to be done

	1	2	3	4	5	6	7	8	9	10	11	12	13	14	15	16	17	18	19
Unit A	Inst Fitter (2)			Welder (2)			Electrician (4)				Welder (3)						Lagger (2)		Recommissioning of all units (1)
Unit B			Mech Fitter (2)		Instrument Fitter (6)						Mechanical Fitter (4)				Welder (4)				
Unit C	Mech Fitter (2)		Electrician (4)								Instrument Fitter (5)					Mech Fitter (3)			
Unit D						Mech Fitter (3)					Electrician (3)			Lagger (3)			Chem Clean (2)		
Unit E	Elec (1)								Mech Fit (1)	Lagger (3)									

Time (days)

(b) Programme of events (for the allocation of resources)

Mechanical Fitter: Unit C (2) | Unit B (2) | Unit D (3) | E (1) | Unit B (4) | Unit C (3)

Electrician: E (1) | Unit C (4) | Unit A (4) | Unit D (3)

Instrument Fitter: Unit A (2) | Unit B (6) | Unit C (5)

Welder: Unit A (2) | Unit A (3) | Unit B (4)

Lagger: Unit E (3) | Unit D (3) | Unit A (2)

Chem Clean Operator: Unit D (2)

Time (days): 1 2 3 4 5 6 7 8 9 10 11 12 13 14 15 16 17 18 19

(c) Time-table for each tradesman

4. For each unit draw a bar chart to a convenient scale and in the correct sequence analysing the work to be done.

   Arrange the respective charts under each other (Fig. 6.1).

5. Draw up a programme of events.

   The object of planning is to arrange a work schedule that will minimize downtime. As a starting point, assume that all the work can be accomplished in the absolute minimum time of 16 days plus 1 day for recommissioning = 17 days. This in effect means that all the work elements of Unit B must be treated as critical and that the work element contained in the other Units (A, C, D and E) must be arranged to fit in with those of Unit B.

   To help visualize and plan the operation refer to Fig. 6.1a, using strips of cardboard of appropriate length to represent the work elements:

   (i) Keep the work elements of Unit B in their present position.

   (ii) Slide (in a horizontal direction only and within the time span of Unit B) the various individual work elements of Units A, C, D and E about until, if possible, a suitable work plan is evolved.

   Note: The respective work sequences of the individual units must not be altered from those originally stated.

   If it is impossible to complete the overhaul in 17 days, then the time span must be increased until the optimum is reached. This in turn might well introduce new, and alter previous, critical activities.

   Figure 6.1b illustrates a suitable work schedule that will give the shortest programme time, i.e. 18 days work plus 1 day recommissioning = 19 days total.

   A closer study of the programme (Fig. 6.1b) will reveal that a number of minor variations are possible in the arrangement of some of the work elements without altering either the overall time span or the basic structure of the programme. In the programme illustrated, we tried to arrange the work within the optimum time span in such a way that as many non-critical elements as possible were provided with 'float' to accommodate operational contingencies that might arise in practice.

6. With information extracted from the work programme a time-table for each tradesman can now be constructed (Fig. 6.1c).

If the time span of the proposed programme must be reduced, the indiscriminate injection of extra resources (tradesmen) or the introduction of overtime working right across the board is unnecessarily wasteful. It is much better to concentrate only on those areas which will yield the best results. To this end the programme (Fig. 6.1b) should be studied to determine:

(a) If the work sequences on the individual units can be changed to bring the work elements closer together, this will also improve the utilization factor of the resources.

(b) If the estimated times of the work element can be reduced by method study, by reduction of the work content. (Initially only the critical activities need receive this attention.)

(c) If the employment of extra tradesmen (resources) or overtime working, for a limited period only, on specific work elements will overcome bottlenecks.

6.2. Fire appliance project

An office block/factory/warehouse complex has a number of fire fighting appliances distributed throughout the site. The maintenance of this equipment is carried out by an outside company of contractors, but it is the responsibility of the site engineer to ensure that the maintenance is executed to a predetermined schedule and programme, and that the work is subsequently recorded.

Devise a system, complete with specimen cards, to ensure that the required work is carried out at the correct time and that the results are recorded for the site engineer's reference. The types of appliances used, together with suggested maintenance schedules are detailed as follows:

6-monthly Maintenance Schedule	
1	Unwind all hose from the reel by taking the nozzle to an open window or other place where a short water discharge will cause no damage
2	Examine hose for bursts or undue wear; check that reel runs freely
3	Ensure that nozzle control is closed, then open reel valve
4	Examine reel and hose for water leaks and remedy as required
5	Open nozzle control, checking for free movement; ensure that water jet is satisfactory
6	Close reel valve, drain water from hose and close nozzle control
7	Rewind hose neatly, wipe over reel, grease as necessary and enter details of check on the maintenance tally

Note. It will normally be found desirable to employ two men on horse-reel maintenance.

Suggested hose-reel maintenance

6-monthly Maintenance Schedule	
1	Examine hose and discharge horn and replace any defective parts
2	Ensure that safety-pin and wire-locking is intact
3	Weigh extinguisher and compare with weight stamped on cylinder neck; if a loss of 5 per cent or more is apparent, return extinguisher to recharging depot
4	Wipe over external surface of unit and enter details of check on the maintenance tally

Note. United Kingdom recharging agents will subject all CO_2 cylinders to a hydraulic pressure test of 23 000 kN/m^2 (3360 p.s.i.) if they have not been so tested within the previous 5 years when presented for recharging. Cylinders are also heated to 150 °C, cleaned, descaled and thoroughly dried. The date of test is then stamped on the neck of the cylinder.

Suggested maintenance for CO_2 extinguishers

Maintenance Schedule	Soda-acid	Foam	Water	Dry Powder
6-monthly* 1 Unscrew and remove head	x	x	x	x
2 Examine and clean relief vent holes, nozzle, plunger and washers	x	x	x	x
3 Examine and clean snifter valve	x			
4 Withdraw acid bottle inner container or CO_2 cartridge	x	x	x	x
5 Stir liquid in outer container with a clean stick	x	x		
6 Agitate powder to ensure it is free-running and has not caked down				x
7 Stir liquid in inner container using a different stick		x		
8 Mix 4 parts from outer container and 1 part from inner container to produce 6 to 8 times the volume of foam mixture: recharge if not satisfactory		x		
9 Ensure that CO_2 cartridge seal is intact			x	x
10 Ensure that acid bottle is intact	x			
11 Lightly grease plunger and check for free movement	x		x	x
12 Lightly Vaseline threads on cap and screw on tightly	x	x	x	x
13 Wipe over external surface of unit and enter details on the maintenance tally	x	x	x	x
3-yearly Discharge extinguisher and re-charge to maker's instructions	x	x	x	x

Note. Each new extinguisher is subjected to a hydraulic pressure test of 2400 kN/m²/(350 p.s.i.) for a period of 5 minutes before leaving the manufacturers. This test should be repeated every 5 years and inexpensive test gear can be purchased for this purpose. Any extinguishers found on any maintenance check to be internally corroded must be withdrawn from service.

Suggested maintenance schedules for commonly-used extinguishers.

Maintenance Schedule
6-monthly 1 Examine for leaks and check liquid level of contents 2 Test pump operation by giving one or two strokes; top up quantity lost during this test 3 Ensure that nozzle is clean 4 Wipe over external surface of unit and enter details of check on the maintenance tally
3-yearly 1 Discharge extinguisher into a clean vessel 2 Lightly grease plunger rod 3 Recharge with original contents, topping up as necessary 4 Wipe over external surface of unit and enter details of check on maintenance tally
Note. Water must never be used for cleaning or flushing carbon-tetrachloride extinguishers, as severe corrosion will result.
Suggested maintenance for carbon-tetrachloride extinguishers

Suggested procedure

1. Inventory – Carry out a physical inventory of all appliances on site.
   Detail: The types.
   Numbers of each type.
   The location of each type.

   The appliances held in the stores as replacements must also be listed.
2. Block plan – On a block plan indicate the position and type of each appliance (Fig. 6.2).
3. Identification numbers – Allocate an individual identification number to each appliance.

   Type and unit number system could be used.

Soda Acid Appliance	01/
Foam	02/
Water	03/
Dry Powder	04/
CO_2	05/
Carbon Tetrachloride	06/
Hose Reels	07/

   Location identification is not required as the appliances will probably rotate.
4. Individual identification – Mark each appliance with its own identification number in a standard manner which is easily visible.
5. Maintenance label – Affix a maintenance label (see Fig. 6.3) to each appliance. (This operation can be carried out when marking the appliance with its identification number.)

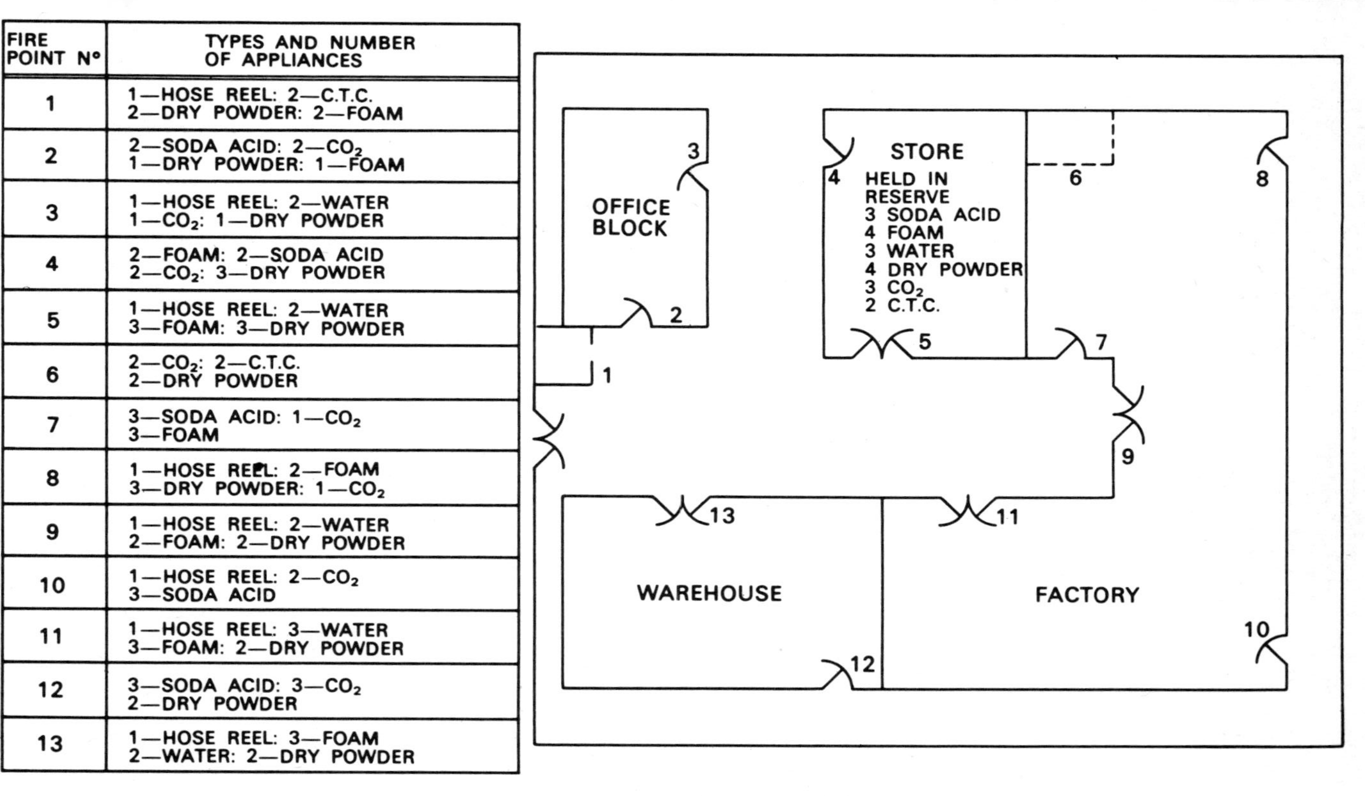

FIRE POINT N°	TYPES AND NUMBER OF APPLIANCES
1	1—HOSE REEL: 2—C.T.C. 2—DRY POWDER: 2—FOAM
2	2—SODA ACID: 2—CO_2 1—DRY POWDER: 1—FOAM
3	1—HOSE REEL: 2—WATER 1—CO_2: 1—DRY POWDER
4	2—FOAM: 2—SODA ACID 2—CO_2: 3—DRY POWDER
5	1—HOSE REEL: 2—WATER 3—FOAM: 3—DRY POWDER
6	2—CO_2: 2—C.T.C. 2—DRY POWDER
7	3—SODA ACID: 1—CO_2 3—FOAM
8	1—HOSE REEL: 2—FOAM 3—DRY POWDER: 1—CO_2
9	1—HOSE REEL: 2—WATER 2—FOAM: 2—DRY POWDER
10	1—HOSE REEL: 2—CO_2 3—SODA ACID
11	1—HOSE REEL: 3—WATER 3—FOAM: 2—DRY POWDER
12	3—SODA ACID: 3—CO_2 2—DRY POWDER
13	1—HOSE REEL: 3—FOAM 2—WATER: 2—DRY POWDER

6.2 Block plan of site. Location of fire fighting appliances

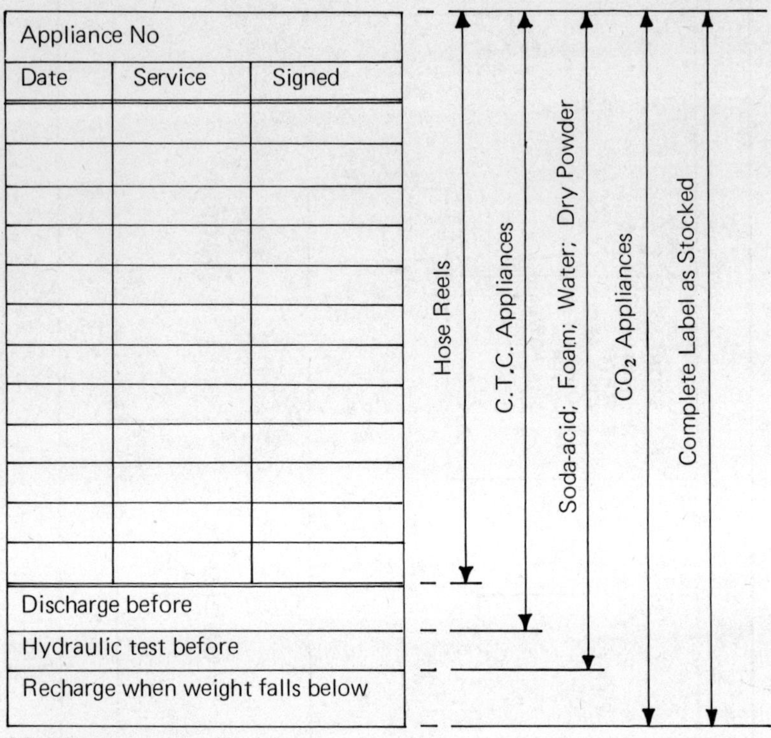

6.3 Maintenance label. To reduce the types of labels in the stores only the complete labels are stocked. The appropriate portion is cut off according to the type of appliance

6. Facility register – Details of each appliance entered in facility register. Type sections – each section has its own colour of cards.
7. History record card – Each appliance to have an individual history record card (see Figs. 6.4(a) to 6.4(d)).

   Different types of appliance – different colour cards, same colour as facility register sections.

   The vertical columns are headed with the date at which maintenance was carried out.

   Note: The list of maintenance/inspection operations has been extracted from the maintenance schedule.
8. Order to contractor – Consult maintenance programme – despatch order to contractor – contractor arrives on site.
9. Documents issued to contractor – When the contractor arrives on site he is issued with:
   (a) Block plan – showing location of appliances.
   (b) Job card – listing the appliances and the service to be carried out. (See Fig. 6.5).

FIRE FIGHTING APPLIANCES INSPECTION/MAINTENANCE RECORD

Date Purchased			Manufacturer	Model	Size
A 6 Monthly B 3 Yearly C 5 Yearly		✓ O.K. X Adjustment Made O Repairs Carried Out			

Service	No.	Item
"A" Service	1	Unscrew and remove head
	2	Examine and clean relief vent holes, nozzle, plunger and washers
	3	Examine and clean snifter valve
	4	Withdraw acid container Ensure acid bottle is intact
	5	Stir liquid in outer container with clean stick
	6	Lightly grease plunger and check for free movement
	7	Lightly grease threads on cap and screw on tightly
	8	Wipe over external surface Check for damage
"B" Service	9	Discharge extinguisher
	10	Examine internally and externally for damage or corrosion
	11	Recharge to manufacturers instructions
"C" Service	12	Hydraulic pressure test 350 lb/sq inch for a period of 5 minutes

19 | 70 | 1 | 2 | 3 | 4 | 5 | 6 | 7 | 8 | 9 | 80 | 1 | 2 | 3 | 4 | 5 | 6 | 7 | 8 | 9 | J | F | M | A | M | J | JY | A | S | O | N | D | Type Soda-acid | Identification No.

6.4 History record cards for fire extinguishers
(a) Soda-acid

FIRE FIGHTING APPLIANCES — INSPECTION/MAINTENANCE RECORD

Date Purchased			Manufacturer	Model	Length
A	6 Monthly	✓ O.K. X Adjustment Made O Repairs Carried Out			
"A" Service	1	Unwind all hose from the reel Check hose runs freely			
	2	Examine for leaks, bursts, or undue wear			
	3	Close nozzle valve, open reel valve Examine for leaks under pressure			
	4	Open nozzle valve, check free movement Ensure water jet is O.K.			
	5	Close reel valve. Drain water from hose. Close nozzle control			
	6	Rewind hose neatly			
	7	Wipe over hose reel			
	8	Grease as necessary			

19 70 1 2 3 4 5 6 7 8 9 80 1 2 3 4 5 6 7 8 9 J F M A M J JY A S O N D — Type Hose Reel — Identification No

(b) Hose reel

FIRE FIGHTING APPLIANCES — INSPECTION/MAINTENANCE RECORD

	Date Purchased		Weight Full Weight Empty CO_2 Contents	Recharge when Wt Falls Below	Manufacturer	Model	Size
	A B	6 Monthly 5 Yearly	✓ O.K. × Adjustment Made O Repairs Carried Out				
"A" Service	1	Examine hose and discharge horn Replace any defective parts					
	2	Ensure that safety-pin and wire-locking are intact					
	3	Check weigh cylinder with weight stamped on cylinder neck					
	4	Return cylinder to be recharged if a loss of 5% or more is apparent					
	5	Wipe over external surface Examine for damage					
"B" Service	6	Discharge contents					
	7	Hydraulic test to 3360 lb/sq in					
	8	Stamp date of test on cylinder neck					
	9	Recharge					

19 | 7/0 | 1 | 2 | 3 | 4 | 5 | 6 | 7 | 8 | 9 | 8/0 | 1 | 2 | 3 | 4 | 5 | 6 | 7 | 8 | 9 | J | F | M | A | M/A | J | J/Y | A/U | S | O | N | D | Type CO_2 Extinguisher | Identification No

(c) CO_2

FIRE FIGHTING APPLIANCES — INSPECTION/MAINTENANCE RECORD

Service	No.	Item	Manufacturer	Model	Size
		Date Purchased			
		A 6 Monthly; B 3 Yearly — ✓ O.K.; X Adjustment Made; O Repairs Carried Out			
"A" Service	1	Examine for leaks			
	2	Check liquid level of contents			
	3	Test pump operation by giving one or two strokes			
	4	Top up quantity lost during test			
	5	Ensure nozzle is clean			
	6	Wipe over external surface of unit			
"B" Service	7	Discharge extinguisher into clean vessel			
	8	Lightly grease plunger rod			
	9	Recharge with original contents, topping up as necessary			
	10	Wipe over external surface of unit			

19 70 1 2 3 4 5 6 7 8 9 80 1 2 3 4 5 6 7 8 9 | J F M A M J JY A S O N D | Type C.T.C. Extinguisher | Identification No.

(d) CTC

Card No	Work Carried Out By:	
Date Issued		
Date Returned	Date Work Carried Out:	
The following fire fighting appliances are to be inspected/maintained as detailed and the results recorded on the attached Job Report Cards.		
Type of Appliance	Service	Identification No of Appliance
Soda-acid	A	
	B	
	C	
Foam	A	
	B	
	C	
Water	A	
	B	
	C	
Dry-powder	A	
	B	
	C	
CO_2	A	
	B	
Carbon-tetrachloride	A	
	B	
Hose Reel	A	

6.5 Job card

(c) Job specification card – (This card is the same as the history record card). One card can indicate the work *to be* carried out on a number of similar appliances on the same visit.

The job specification card is also used as the

Job report card – one card can indicate the work *actually* carried out on a number of similar appliances on the same visit.

In this case the vertical columns are headed with the appliance identification number.

10. Work is carried out – The contractor carries out the work to the appliances listed on the job card, and records of the work carried out on the job report card.

    The contractor also completes the maintenance label attached to the appliance.

11. Recording information – The contractor gives the completed job report cards and job card into the maintenance office/works engineer.

    Appliances *actually* inspected/maintained are checked against appliances listed on the job card.

    Information detailed on the job report card is transferred to the respective history card.

12. Maintenance programme – Adjust programme, move date indicator to next date when inspection/maintenance is to be carried out.

    Red marker – Indicates 'A' Service.

    Blue marker – Indicates 'B' Service.

    Green marker – Indicates 'C' Service.

    The base of the history record card has years 1970 to 1989 and also months of the year January to December printed along the edge.

    An appropriate coloured marked (a red, or a blue, or a green) in the month section indicates the type (A, B or C) of the next service and the month in the current year it is to be carried out.

    A blue marked and a green marked in the year section indicates the year the next 'B' and 'C' service is to be carried out.

6.3. Plant availability project

A large company operates several modern continuous process plants in different parts of the country. Each plant is similar and manufactures industrial chemicals. Records show that the average overall production time of these plants is 85 per cent, whereas, each plant was planned to produce for 93 per cent of the possible time. (Possible time 168 hours per week : i.e. 168 hours = 100 per cent.)

Draw up a general overall plan to bring the production time of these plants up to the required target of 93 per cent.

Additional information:

The process machinery at each plant is modern, well-constructed, heavy-engineering equipment.

It includes:

Compressors (air and gas); Heat exchangers; Cooling water systems (cooling towers, etc.); Pressure vessels; Pumps (water and chemical); Reaction vessels; Storage tanks and hoppers; Pipe lines;

Instrumentation; Electric motors of all sizes including their attendant control and switchgear.

The plants operate – 24 hours per day: 7 days per week: 52 weeks per year: Every two years there is a planned three week shut down period for the complete overhaul of the plant.

The production staff complete a daily plant log sheet which records all the relevant hourly readings and operating conditions. This sheet also records all plant and production stoppages, breakdowns and failures giving the reason for each and the resultant downtime. The log sheets are filed for a period of two years after which they are destroyed.

Each plant has a team of skilled conscientious tradesmen to carry out the normal day to day maintenance. They are assisted by the production personnel if the need arises.

During the overhaul, manufacturers' engineers and outside contractors are called in to carry out specialist work.

No comprehensive planned maintenance system exists, maintenance is arranged on a short term basis either according to the operational hours of the plant or when it is considered necessary. The plant is maintained in good condition. Running maintenance is usually carried out within the recommended period.

The nature of the process necessitates occasional shut down to clear lines, wash out vessels, etc. Frequently, shut down maintenance is arranged to coincide with these periods. Otherwise, shut downs are arranged solely to carry out this maintenance.

When breakdowns occur, every effort is made to ensure that the repair is carried out in the minimum time, as the longer the plant is shut down and the process stopped the longer it takes the plant to regain operating conditions and begin producing again. When shutting down or stopping the plant a routine procedure must be adhered to (i.e. depressurize, drain down, etc.). The time required to carry out these operations depends upon various process conditions and factors.

Adequate stocks of most of the essential spares are held in the stores on a stock controlled basis.

There is a steady rise in customer demand for the company's products. This demand can only be met by full production at the target figures. It is estimated that if production hours reached the target figure there would be a profit increase of approximately £40 000 per annum.

Suggested procedure

Modern, high-output, continuous process plant requires heavy capital expenditure and is expected to achieve a high rate of return upon the investment made, thus its high utilization is an economic necessity. Downtime is extremely expensive not only in terms of idle capital but also in terms of lost production and revenue.

Most large companies are aware of the total cost of downtime and are continually endeavouring to reduce it, often by employing full-time specialists. Small companies may not have the resources or consider that it is justified financially to go to such lengths, but prefer to attempt the solution of each problem on a piecemeal basis as and when it arises. This attitude is not the prerogative of only the small companies. Rarely is the problem studied overall or analysed logically in sufficient detail to provide the most effective long-term solution.

Although most managements are conscious of the need to reduce downtime, they often cannot determine the most effective and economical method to accomplish it. Local management is usually more than occupied with the day to day problems that arise and may not be able to devote either sufficient time or qualified staff to carry out a thorough investigation, and to formulate its own comprehensive plan from scratch. Again, local management may be too near the problem to see the wood for the trees.

There is no universal solution for every case, each plan must be tailored to suit the circumstances, but there are certain fundamental principles that are common to all. These principles when expressed in a practical and easily applied manner provide the basis of a general method for setting up a programme to reduce downtime.

The average company does not have usually the services of mathematicians, computers, programmers, and an abundance of highly qualified staff available to formulate, instal and operate such systems. Therefore, if a plan is to be acceptable and workable it must be presented in a clear and simple form so that it is useable but not difficult to implement or operate by persons of average intelligence. It should require the minimum of expensive equipment, paperwork, recording and staff to operate it, but must be capable of supplying the required information quickly, accurately and in a form that is understood easily. With these conditions in mind the problem was considered and the following approach worked out. Although the investigation and analysis described in the text are concerned directly with the particular case quoted, the general approach and procedure could be adapted easily to suit many other similar situations.

Any project will be accomplished more effectively if it follows a carefully pre-determined plan. Such a plan provides a systematic means for guiding all activities in the most effective and constructive way, by ensuring that productive effort does not deviate, or is sidetracked, from the objective. Although, initially, it may not be feasible to plan down to the very last detail, it should be possible to set out the broad outlines indicating the method of approach so that the work may proceed in a progressive logical order.

The procedure set out below is a summary of the approach:

1. Define the objective

What is the problem?
What is the ultimate objective?

2. Establish the facts
   What are the present facts?
   What is done now?
   How is it done now?
   What happens now?
3. Examine the facts
   Analyse the facts critically and systematically.
4. Determine the cause
   What is the root of the trouble?
   What are the contributing factors?
   What are the individual and collective effects of each?
   Why, how and when does it happen?
5. Determine the remedy
   What is the *best* plan?
   What must be done?
   How and when (short, medium and long-term) must it be done?
6. Apply the remedy
7. Check the results
   Compare the actual results with previous results and with the target.
   Determine where controls must be applied.
8. Apply control
   Apply control when and where necessary.

Each of the above stages will now be considered and discussed in detail.

1. *Define the objective*
   (a) What is the problem? – to reduce downtime.
   (b) What is the ultimate objective? – to attain and maintain plant availability of 93 per cent within a reasonable period of time and at an economical cost.
2. *Establish the facts*

Management is aware usually of the extent of downtime but frequently does not know the contributing factors or the effect of each. It is only when these facts are known and the size of the problem assessed that it is possible to plan in detail the best course of action. The more facts that can be established the better will be the diagnosis and remedy.

To arrive at comparable results, each plant must be examined identically. Also, to obtain a reasonably accurate picture of events, operating conditions, types of breakdown, and stoppages and failures, the period under examination should be as long as is practicable. In the present case, it would be possible and also desirable to cover a period of 2 years, as:

(a) the log sheets are available for this period of time, and
(b) this period will include a plant overhaul.

It was previously stated that 'the average overall production time of these plants is 85 per cent, the production time of some plants will be

		Target	Plant No 1	Plant No 2	Plant No 3
A	Period under review	2 years = 104 weeks	2 years = 104 weeks	2 years = 104 weeks	2 years = 104 weeks
B	Total possible hours	104 x 168 = 17,472 hrs	104 x 168 = 17,472 hrs	104 x 168 = 17,472 hrs	104 x 168 = 17,472 hrs
C	Total actual production hours obtained from log sheets	16,250 hrs	14,850 hrs	14,600 hrs	15,150 hrs
D	Total downtime hours B–C	1,222 hrs	2,622 hrs	2,872 hrs	2,322 hrs
E	Percentage downtime $\frac{D}{B} \times 100$	7%	15%	16·4%	13·3%

Allocation of Downtime Hours

F	Total downtime hours due to overhaul period	3 weeks 3 x 168 = 504 hr	3 weeks 3 x 168 = 504 hr	3 weeks 3 x 168 = 504 hr	3 weeks 3 x 168 = 504 hr
G	Total downtime hours to be spread over remaining period (D–F)	718 hrs	2118 hrs	2368 hrs	1818 hrs
H	Average weekly downtime over remaining period (G ÷ 101)	7·1 hrs	19·2 hrs	23·5 hrs	18 hrs

6.6 Chart no. 1

higher than this, while others are lower. The downtime of each plant must be determined so that we can compare the various plants and assess the deviation of each from the ultimate target. Figure 6.6 sets out this information in tabular form (specimen data have been used to illustrate more effectively the case in point).

The root causes of downtime are numerous and varied, but without exception they can be classified into a set number of categories which are common to all – only their magnitudes and relative proportions are different. To determine the magnitude and relative importance of these common factors, the total downtime of each plant must be analysed into main categories in sufficient detail to enable the frequency and contribution of each downtime element to be determined and tabulated. These categories are listed in Fig. 6.7, the relevant data from each plant being entered upon separate, identical charts. (For the purpose of the example, numerical values assumed to originate from Plant No. 1 have been used in subsequent calculations.)

When this form of analysis is used for other types of plant or industry, the categories shown in Fig. 6.7 would be adapted to suit the individual circumstances, but in general they would follow closely those shown in the example.

To show the pattern of occurrence of each downtime element as well as the pattern of total weekly downtime hours, a 'Programme of past events', Fig. 6.8, was reconstructed from information contained in log sheets, works records and other available sources. A graphical representative of the total weekly downtime hours is shown in Fig. 6.9.

In a programmed stop, a number of different tasks, both of an engineering and a process nature, may be carried out simultaneously. This can make it difficult to allocate the hours of downtime to any particular element for analytical purposes. In these situations, downtime can be allocated, either by direct proportion of the work content of the various elements or by a method of priorities – the element having the highest priority bears the largest allocation of downtime hours. Whatever method is eventually used it should be applied consistently throughout the investigation.

3. *Examine the facts*

With the aid of the charts in Figs. 6.7 to 6.9 it is possible to obtain a reasonably accurate picture of the pattern, timing and reasons for the various stoppages. This overall analysis will highlight areas requiring further examination.

Check list (to assist investigation):

What is the age of the plant?
What is the material condition of the plant?
Are all the stoppages necessary?
What is the maximum length of time for which each piece of equipment can operate before requiring shut-down maintenance?

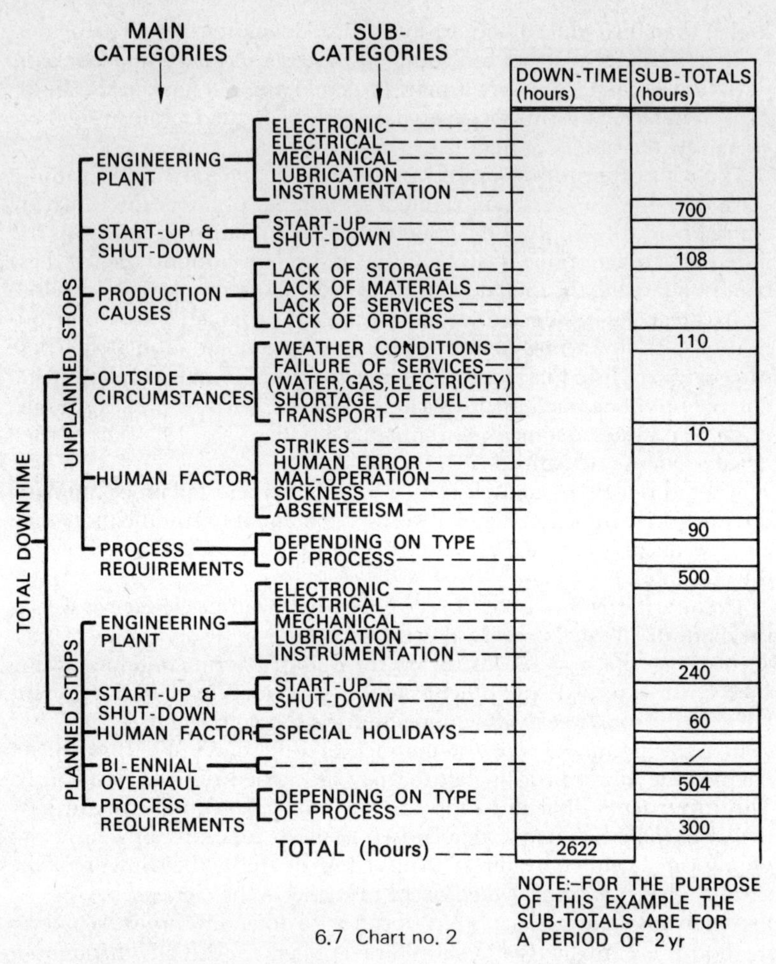

6.7 Chart no. 2

Is the plant overloaded?
What is the standard of maintenance?
Under the present circumstances, is sufficient manpower available to cope with the maintenance load?
How long does it take the maintenance personnel to answer a call?
Is there a planned maintenance scheme in operation?
(a) To what extent?
(b) How far ahead is planned?
(c) Do all departments co-operate fully (i.e. production and maintenance departments)?
(d) Is it effectively controlled?
What is the operational standard of the production personnel?

	WEEK N°	1	2	3	4	5	6	7	8	100	101	102	103	104
UNPLANNED STOPS	ENGINEERING PLANT	6	5	7	1	7	3	2	8					
	START-UP & SHUT-DOWN	2	1	2	1	2	1	1	2					
	PRODUCTION CAUSES	2	1	/	/	2	/	/	5					
	OUTSIDE CIRCUMSTANCES	/	/	/	/	2	/	/	/					
	HUMAN FACTOR	1	/	/	/	/	2	/	/					
	PROCESS REQUIREMENTS	5	2	9	3	5	4	3	3					
PLANNED STOPS	ENGINEERING PLANT	/	4	/	3	3	/	3	3					
	START-UP & SHUT-DOWN	1	1	/	1	1	/	1	1					
	HUMAN FACTOR	/	/	/	/	/	/	/	/					
	BI-ENNIAL OVERHAUL	/	/	/	/	/	/	/	/					
	PROCESS REQUIREMENTS	3	1	/	3	3	/	/	4					
	TOTAL WEEKLY DOWN-TIME (hours)	20	15	18	12	25	10	14	26					

6.8 Chart no. 3. Programme of past events

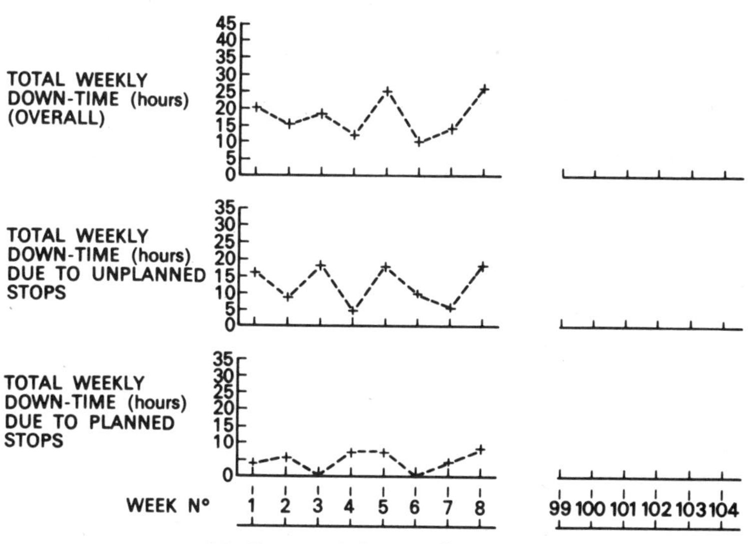

6.9 Chart no. 4. Pattern of past events

4. *Determine the cause*

By sifting through the facts, we can identify the causes of downtime and its prolongation. Some of the more common causes can result from:

Failure through fair wear and tear.
Unreliable equipment: design faults
material faults.
Insufficient maintenance: poor workmanship
incorrect maintenance
infrequent maintenance
absence of maintenance.
Incorrect or insufficient lubrication.
Lack of spares.
Lack of tools or specialized equipment.
No planned maintenance system.
Incorrect maintenance planning:
(a) too often – causing unnecessary shutdown.
(b) not often enough – causing premature or excessive failure.
(c) lack of co-operation between departments – shut downs for engineering maintenance and for process requirements should be planned to coincide.
(d) incorrect use of plant and equipment by production staff.

By plotting the various elements shown in Fig. 6.7 against their overall effect, (Pareto Curve, Fig. 6.10) we identify the significant factors accounting for the downtime. In this particular example, the Pareto Curve indicates that a major portion of the downtime is caused by three elements:

Unscheduled stops – engineering plant	700 hrs.
Plant overhaul	504 hrs.
Unscheduled process requirements	500 hrs.

and to a lesser extent:

Scheduled process requirements	300 hrs.
Unscheduled stops – engineering plant	240 hrs.

Having determined the main downtime elements, the contributory or root causes of these elements must be found.

5. *Determine the remedy*

The remedy will depend upon the diagnosis which will differ in each case, but in this particular example one of the remedial actions might well be tighter planning of the overhaul operation to reduce its time span.

Any proposed remedy must be an economic proposition capable of proving itself within a reasonable period of time. It is often convenient to consider it in three distinct phases:

(a) Short term – work or procedures that are inexpensive, easily implemented, not controversial and capable of producing good, quick results.

ELEMENT	ELEMENTAL VALUE (hours)	CUMULATIVE VALUE OF ELEMENTS	PERCENTAGE EFFECT	PERCENTAGE OF ELEMENTS
UNPLANNED ENG PLANT	700	700	26·7	9·1
PLANT OVERHAUL	504	1204	45·9	18·2
UNPLANNED PROCESS	500	1704	65·0	27·3
PLANNED PROCESS	300	2004	76·5	36·4
PLANNED ENG PLANT	240	2244	85·5	45·45
UNPLANNED PRODN. CAUSES	110	2354	90·0	54·5
UNPLANNED S/U—S/D	108	2462	94·0	63·6
UNPLANNED HUMAN FACTOR	90	2552	97·5	72·65
PLANNED S/U—S/D	60	2612	99·5	81·7
UNPLANNED O/S CIRCUM:	10	2622	100·	90·8
PLANNED HUMAN FACTOR	0	2622	100·	100·

NOTE:– % ELEMENT = $\frac{100}{\text{N° OF ELEMENTS}} = \frac{100}{11}$ = 9·1% PER ELEMENT

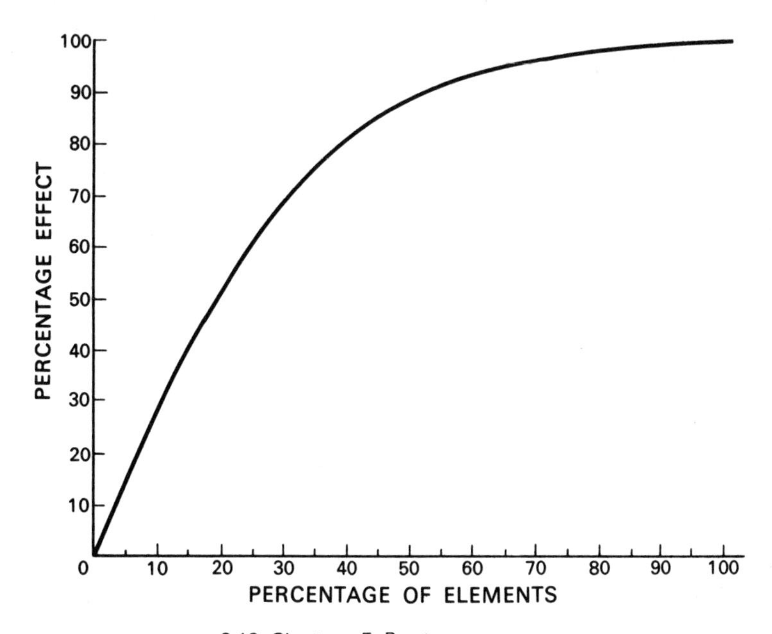

6.10 Chart no. 5. Pareto curve

(b) Medium term – the establishment of interrelated production and maintenance programmes.
Planned maintenance.
Training schemes – for maintenance and production staff.
Stock control of spares, etc.

(c) Long term – research and design into processes, machines, materials, methods and maintenance.

6. *Apply the remedy*

The actual method of applying the remedy will vary from company to company (and even from factory to factory within the same company) and will depend upon corporate and local policies.

7. *Check the results*

The results must be checked periodically to ensure they are attaining or approaching those anticipated. Again, charts similar to Figs. 6.7, 6.8 and 6.9 should be used for this purpose. Chart 2 (Fig. 6.7) being completed each week, Charts 3 (Fig. 6.8) and 4 (Fig. 6.9) compiled progressively over two years.

The construction of a 'Z' Chart (Fig. 6.11) showing a moving biennial total of downtime will illustrate the progress.

Note: as availability improves, so it becomes more and more difficult as well as costly to make further gains – it follows the law of diminishing returns.

TOTAL WEEKLY DOWN-TIME (hours)	CUMULATIVE WEEKLY TOTAL OF DOWN-TIME (hours)	MOVING BI-ENNIAL TOTAL OF DOWN-TIME (hours)	MOVING BI-ENNIAL PERCENTAGE OF DOWN-TIME (%)
20	20	2622	15
19	39	2611	15
15	54	2602	14·95
10	64	2592	14·9
13	77	2580	14·8
14	91	2570	14·75
12	103	2561	14·7
11	114	2552	14·65
10	124	2542	14·6
12	136	2530	14·55

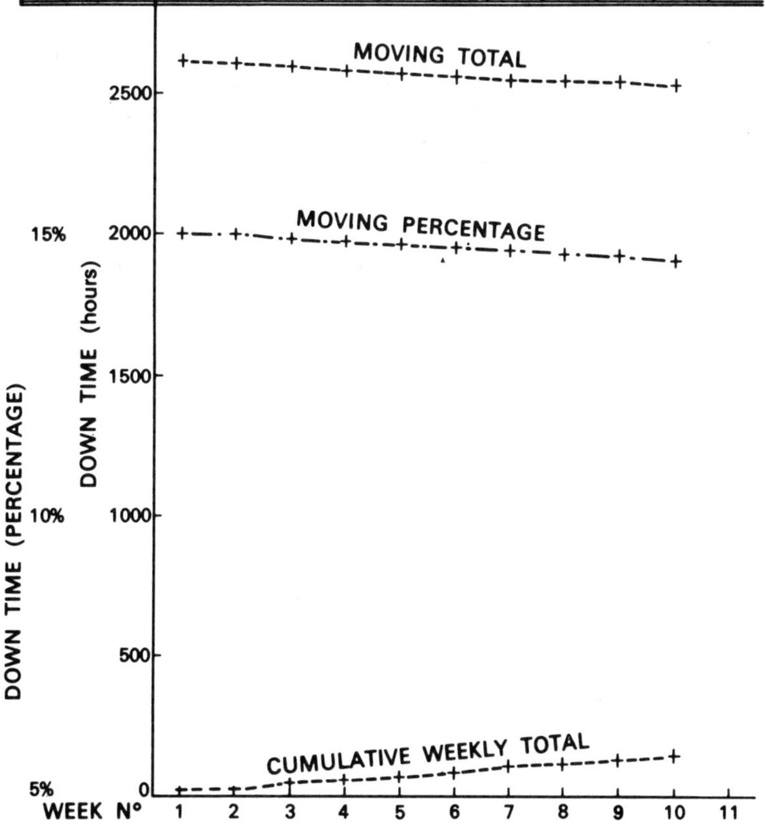

MOVING TOTAL
MOVING PERCENTAGE
CUMULATIVE WEEKLY TOTAL
DOWN TIME (PERCENTAGE)
DOWN TIME (hours)
15% 10% 5%
2500 2000 1500 1000 500 0
WEEK N° 1 2 3 4 5 6 7 8 9 10 11

6.11 Chart no. 6. Z chart

6.4. Factories Acts project

A minor incident in a factory drew the attention of senior management to serious shortcomings in the application of the Factories Acts. A subsequent investigation revealed, among other things, instances of equipment not being examined or maintained in accordance with the regulations; of incomplete records; of items being overlooked or omitted; of difficulty relating equipment to its relevant documentation.

In view of the serious legal and safety implications involved the need to rectify the situation was *urgent*.

(a) Draw up a plan of action to quickly correct the situation, and

(b) Formulate a system to ensure future compliance with the Acts.

Note: With certain exceptions, it is the occupier of the factory who is responsible for seeing that the law, in respect of the Factories Acts, is complied with. He cannot delegate his responsibility to an employee, a specialist organization or a contractor. Figure 6.12 is an example of a document designed to assist management to comply with the Factories Acts.

TEST AND EXAMINATION OF CHAINS, ROPES OR LIFTING TACKLE (EXCEPTING FIBRE ROPES OR FIBRE ROPE SLINGS) BEFORE BEING TAKEN INTO USE AND DURING USE IN THE FACTORY

WORKS NO.	DESCRIPTION OF CHAIN, ROPE OR LIFTING TACKLE	DATE TAKEN INTO USE	TEST LOAD	S.W. LOAD

DATE OF CERTIFICATE OR EXAMINATION	CERTIFICATE NO. OR SIGNATURE OF EXAMINER	NAME AND ADDRESS OF PERSON WHO ISSUED THE CERTIFICATE	PARTICULARS OF ANY DEFECT FOUND AFFECTING THE SAFE WORKING LOAD	STEPS TAKEN TO REMEDY SUCH DEFECTS	DATE OF ANNEALING

Kalamazoo 213137-001

TITLE BELOW THIS LINE

WORKS NUMBER	MAKERS REFERENCE	LOCATION	EXAMINATION	ANNEALING

6.12 A document to assist factory management comply with the Factories Acts (Kalamazoo).

Appendix 1

General terms used in maintenance organization

British Standard 3811 : 1964

Glossary

No.	Term	Definition
01	maintenance	Work undertaken in order to keep or restore every facility, i.e. every part of a site, building and contents, to an acceptable standard. NOTE 1. Where there are statutory requirements for maintenance the 'acceptable standard' to be reached must be not less than that necessary to meet such requirements. NOTE 2. 'Maintained' is defined in the Factories Act, 1961 as: 'Maintained in an efficient state, in efficient working order and in good repair'.
02	availability	Period during which a facility is in a usable condition.
03	availability factor	$\frac{\text{Availability}}{\text{The period of maximum requirement}}$
04	breakdown maintenance	(This means planned corrective breakdown maintenance.) Work which is carried out after a failure, but for which advance provision has been made, in the form of spares, materials, labour and equipment.
05	check	To compare with an acceptable standard by suitable or defined means, whilst the facility is non-operational.
06	clean	To reduce contamination to an acceptable standard.
07	contract maintenance	Work carried out by a contractor.
08	corrective maintenance	(This means planned corrective maintenance.) Work undertaken to restore a facility to an acceptable standard.

09	down time	Period during which a facility is not ready for use.
10	emergency maintenance	(This means unplanned emergency maintenance.) Work necessitated by unforseen breakdowns or damage.
11	facility register	A record of facilities, including information such as constructional and technical details about each. This may be combined with an inventory.
12	fault report	A document reporting departure of a facility from an acceptable standard.
13	history card	A document on which information about all work done on and/or by a particular facility is recorded.
14	inspection	The process of ensuring by assessment that a facility reaches the necessary standard of quality or performance and that the level is maintained.
15	inventory	A list of all facilities, i.e. all parts of a site, building and contents, for purposes of identification.
16	job interval	The period, which may be based on time or other factors, between occurrences of a particular maintenance job.
17	job report	A statement recording the work done and the condition of the facility.
18	job card	(Non-preferred term.)
19	job specification	A document describing the work to be done.
20	maintenance management	The organization of maintenance within an agreed policy.
21	maintenance planning	Deciding in advance the jobs, methods, materials, tools, machines, labour, timing and time required.
22	maintenance programme	A list allocating specific maintenance to a specific period.
23	maintenance programming	The preparation of a maintenance programme.
24	maintenance schedule	A comprehensive list of maintenance and its incidence.
25	maintenance scheduling	The preparation of a maintenance schedule.
26	outage, planned and unplanned	(Non-preferred term.
27	overhaul RECONDITION REFIT REBUILD	A comprehensive examination and restoration of a facility, or a major part thereof, to an acceptable standard.
28	permit to work	A signed document, authorizing access to a facility, which must define conditions, including safety precautions, under which work may be carried out.

28 (continued)		This may include a document, signed on completion of maintenance, stating that a facility is safe and ready for use.
29	planned maintenance	Work organized and carried out with forethought, control and records.
30	preventive maintenance*	(This means planned preventive maintenance.) Work which is directed to the prevention of failure of a facility.
31	routine maintenance*	(Non-preferred term. See Appendix B.)
32	running maintenance	(This means planned preventive running maintenance.) Work which can be carried out whilst the facility is in service.
33	Scheduled maintenance*	(Non-preferred term. See Appendix B.)
34	servicing SERVICE	(Non-preferred term. See Appendix B.)
35	shut-down maintenance	(This means planned preventive or corrective shut-down maintenance.) Work which can only be carried out when the facility is, or is taken, out of service.
36	test	To compare with an acceptable standard by suitable or defined means, whilst the facility is operational.
37	user	Anyone for whom a facility performs its function.
38	utilization factor	$\frac{\text{Availability}}{\text{The period of actual use}}$
39	work order	A written instruction detailing work to be carried out.
40	work requisition	A document requesting work to be carried out.

Non-preferred terms

B18	job card	A synonym for work order
B26	outage, planned and unplanned	A term used in certain industries, e.g. electricity supply, to denote when a facility is not in use or is not required for use. In the context of maintenance it is a synonym for 'down time'.
B31	routine maintenance*	A term sometimes used in the context of maintenance tasks of a repetitive and minor nature. It should not be used, as this type of work is part of that covered by the term 'running maintenance' and 'shut-down maintenance'.

B33	scheduled maintenance•	A term sometimes used to define the fields covered by 'running maintenance' and 'shut-down maintenance'.
B34	servicing SERVICE	A term sometimes used to describe the act of carrying out minor planned maintenance.

•For the special application of these terms to automatic data processing only see B.S. 3527, 'Glossary of terms relating to automatic data processing'.

Appendix 2

The Factories Acts

It is the plant engineer's responsibility to ensure the availability and efficiency of existing plant and buildings; equally, he has a moral responsibility to ensure that the equipment and its operation are not hazardous. This moral responsibility is enforced, and reinforced by Statutory Regulations – viz. The Factories Acts. These Acts have been specially framed to safeguard the employee's welfare, safety and health.

At first sight, any direct connection between the Factory Acts and planned maintenance may seem rather remote until it is realized that the Acts require:

(a) That certain classes of plant and equipment must receive periodic attention (examination, inspection, etc.).

(b) That these examinations must be carried out within stated intervals of time.

(c) That the reports of the results of these examinations must be made available for future reference.

(d) The equipment must bear a distinguishing mark so that it may be positively identified.

The Acts take on an obligatory form of preventive maintenance which in this case is directed towards those facilities whose failure could create a hazard. Failure to comply fully with the regulations could lead to serious consequences, therefore a means of scheduling and programming the work is essential.

Equipment that is part of the process, i.e. air compressors, receivers, boilers, etc. would be included within the overall maintenance programme, while individual items – chains, slings, etc. – may be programmed separately.

As a reminder, and to assist the programme planner in incorporating these statutory requirements into his programme, extracts from the Acts concerning equipment most common to factories are listed in the following pages.

Note: Special regulations within the Acts apply to the building and construction industries, shipbuilding and ship repairing industries, docks and harbours. These differ slightly from those detailed and the relevant sections have to be consulted.

FACILITY (EQUIPMENT OR PLANT CONCERNED)	STATUTORY INTERVAL OF ATTENTION	FACTORY ACT REFERENCE		SCHEDULE OF REQUIREMENTS
		Factories Acts *Part 1*		
Factories Cleanliness	Weekly	Sec 1	: s/s 2b	The floor of every workroom should be cleaned at least *once every week* by washing, or, if effective and suitable, by sweeping or other method.
			: s/s 3	The following provision shall apply as respects all inside walls and partitions, and all ceilings or tops of rooms and all walls, sides and tops of passages and stair ways:
	14 months		: s/s 3a	Where they have a smooth impervious surface they shall at least *once* in every period of *14 months* be washed with hot water and soap or other suitable detergent or cleaned by such other method as may be approved by the inspector for the district.
	7 years 14 months		: s/s 3b	Where they are kept painted in a prescribed manner or varnished, they shall be repainted in a prescribed manner, or revarnished at such intervals of *not more than 7 years* as may be prescribed, and shall at least *once* in every period of *14 months* be washed with hot water and soap or other suitable detergent or cleaned by such other method as may be approved by the inspector for the district.
	14 months		: s/s 3c	In any other case they shall be kept whitewashed or colour washed and the white washing or colour washing shall be repeated at least *once* in every period of *14 months*. *Note:* Certain premises are exempted from the above orders, (consult the Acts regarding the exemption) but the exemption does *not* apply to Maintenance Shops, Messrooms, Cloakrooms, Lavatories and Sanitory Conveniences of those premises.
		Part 2		
Hoists and Lifts (General) (Consult F.A. Re Definition within the Act)	6 months	Sec 22	: s/s 2	Every hoist or lift shall be thoroughly examined by a competent person at least *once* in every period of *6 months*.

Chains, Ropes and Lifting Tackle	6 months	Sec 26	: s/s 1d	All chains, ropes and lifting tackle in use shall be thoroughly examined by a competent person at least *once* in every period of *6 months* or at such greater intervals as the Minister may prescribe.
	14 months 6 months		: s/s 1f	Every chain and lifting tackle, except a rope sling shall, unless of a class or description exempted by certificate of the Chief Inspector upon the ground that it is made of such material or so constructed that it cannot be subjected to heat treatment without risk of damage or that it has been subjected to some form of heat treatment (other than annealing) approved by him, be annealed at least *once* in every *14 months*, or in the case of chains and slings of $\frac{1}{2}$ *inch* bar or smaller, or chain used in connnection with molten metal or molten slag in every *6 months*, except that chains and lifting tackle not in regular use need be annealed only when necessary.
Cranes and Other Lifting Machines	14 months	Sec 27	: s/s 2	All such parts and gear shall be examined by a competent person at least *once* in every period of *14 months*.
Dangerous Fumes. (Breathing Apparatus)	Monthly	Sec 30	: s/s 6	Any factory in which work has to be done inside a chamber, vat, pit, pipe, flue or similar confined space in which dangerous fumes are liable to be present to such an extent as to involve risk of persons being overcome thereby, there shall be provided and kept readily available a sufficient supply of breathing apparatus of a type approved by the Chief Inspector, of belts and ropes, and of suitable reviving apparatus and oxygen, and the apparatus, belts and ropes shall be maintained and thoroughly examined at least *once a month* or at such other intervals as may be prescribed by a competent person and a report on every such examination, signed by the person making the examination and containing the prescribed particulars shall be kept available for inspection.
Steam Boilers (Maintenance Examination & Use)	14 months 26 months	Sec 33 Exam. of Steam Boiler Regs. Reg. 4		All steam boilers with the exception of those detailed below shall be examined at least *once* in every period of *14 months*. The following intervals, after the initial 14 months of re-examination apply to steam boilers of any of the following kinds in the case of which a period of 21 years has not expired since it was first taken into use. (a) Water tube boiler of which the drum and any headers are of fusion welded or solid forged construction which has an evaporative capacity of not less than 50 000 lb of steam per hour. *26 months*

FACILITY (EQUIPMENT OR PLANT CONCERNED)	STATUTORY INTERVAL OF ATTENTION	FACTORY ACT REFERENCE		SCHEDULE OF REQUIREMENTS
	26 months	Factories Acts *Part 2*		(b) A boiler in a group of water tube boilers of which the drums and any headers are of fusion welded or solid forged construction being a group in which: (i) Each boiler has an evaporative capacity of not less than 25 000 lb of steam per hour, and (ii) The total evaporative capacity of all the boilers is not less than 100 000 lb of steam per hour. *26 months*
	26 months			(c) A boiler which is a waste heat boiler or heat exchanger with fusion welded logitudinal and circumferential seams, or a superheater of fusion welded construction, and which is an integral part of a continuous flow installation, in a chemical or oil refinery processing plant. *26 months*
Steam Receivers & Containers	26 months	Sec 35	: s/s 5	With certain specific exceptions – see exemption list, shall be examined so far as the construction of the receiver permits, at least *once* in every period of *26 months*.
Air Receivers	26 months	Sec 36	: s/s 4	Every Air Receiver shall be thoroughly cleaned and examined at least *once* in every period of *26 months*. (In certain circumstances the examination of solid drawn receivers may be extended to within a period of *4 years* but cleaning must still take place within a period of *26 months*.
Water-Sealed Gas Holders	2 years	Sec 39	: s/s 2	Water-sealed gas holders which have a storage capacity of not less than 5000 *cubic feet*. Shall be thoroughly examined externally by a competent person at least *once* in every period of *2 years*.
	10 years		: s/s 3	In the case of a gas holder of which any lift has been in use for more than 20 years, the internal state of the sheeting shall, at least *once* in every period of *10 years*, be examined by a competent person by cutting samples from the crown and sides of the holder and make a report on every such examination.

Fire Warnings	3 months	Sec 51 : s/s 1	There shall be tested or examined at least *once* in every period of *3 months* and whenever an inspector so requires every means of giving warning in case of fire which is required to be provided by or under this Act.
Electricity Insulating Stands	As necessary	Electricity Regulations Reg. 23	Where necessary adequately to prevent damage, insulating stands or screens shall be provided and kept permanently in position, and shall be *maintained in sound condition*.
	As necessary	Reg. 24	Portable insulating stands, screens, boots, gloves, or other suitable means shall be provided and used when necessary adequately to prevent damage, and shall be *periodically* examined by an authorized person.
Grinding of Metals (Ventilation)	6 months	Grinding of Metals Regs. Reg. 17(a)	All ventilating plant used for the purpose of extracting or suppressing dust shall at least *once* in every *6 months* be examined and tested by a competent person, and any defect disclosed by such examination and test shall be rectified as soon as possible.
		Reg. 17(b)	A register (Form 89) containing particulars of such examination and test shall be kept in a form approved by the Chief Inspector and be available for inspection.
Power Presses		Power Presses Regulations Reg. 5: 2	No power press shall be used unless it has been thoroughly examined and tested by a competent person.
	12 months	(a)	In the case of a power press on which the tools are fenced exclusively by means of fixed fencing within the immediately preceding period of *12 months* or in any other
	6 months	(b)	case, within the immediately preceding period of *6 months*.
	6 months	3	No power press shall be used unless every safety device (other than fixed fencing) thereon has within the immediately preceding period of *6 months* when in position on that power press, been thoroughly examined and tested by a competent person.
Radiation Sources		The Ionizing Radiations (Sealed Sources Regulations) Reg. 15 (2)	A distinguishing number or other identifying mark shall be on or attached to every sealed source.
	26 months	(3)	The prescribed test for leakage of radioactive substance shall be made by a qualified person at least *once* in every period of *26 months*.

Appendix 3

Questions from examination papers

The following questions were taken from past examination papers of City and Guilds of London Institute, Mechanical Engineering Technicians Course No. 255.

Section A contains questions from Part II – Plant Maintenance and Works Services.

Section B contains questions from Part III – Plant Engineering.

Section A

1. List the various plant records that should be kept by a works engineer, outlining what service he may obtain from the information recorded. (1965)

2. (a) Why is stock control necessary?
   (b) Briefly describe an effective stock control system.
   (c) State four factors that govern the numbers of spares held for any particular item of plant. (1966)

3. (a) Specify three items that should appear on an operators daily check for a fork lift truck.
   (b) Draw up a typical maintenance programme for a fork lift truck. Time intervals based on running hours. (1968)

4. Lifting chains and steel wire ropes have to be tested and examined to ensure safety and to satisfy statutory requirements.
   Draw up a combined inventory card and examination schedule for a steel-wire rope sling *or* a steel chain sling stating the type and size of chain. (1969)

5. A number of electrically driven pumps of similar type are to be installed in a pump house. What records would you suggest be made and maintained for either:
   (i) The pumps (state type), *or*
   (ii) The electrical equipment (state type).
   Give reasons for your choice of records. (1969)

6. The maintenance and service department of a factory is to be made responsible for organizing the initial and continued purchase storage and control of all materials and spares required to maintain and service the plant.

(a) State six factors that should be considered in the purchase of spares.
(b) Describe a system of controlling the availability and issue of spares and explain why such control is desirable. (1969)

7. (a) Explain the term 'planned preventive maintenance'.
(b) Give particulars of a planned preventive maintenance system suitable for *one* of the following:
(i) a large machine shop
(ii) an electricity generating plant
(iii) a battery of air compressors.
(c) Explain how to deal with an unexpected breakdown in the plant you have chosen. (1972)

Section B

1. (a) Summarize five main aims of maintenance operations in any factory. (Not more than a single short sentence is required for each aim.)
(b) What records may be used to indicate the effectiveness of the maintenance function?
(c) How should these records be analysed and presented? (1968)

2. (a) What is meant by the terms:
(i) scheduled maintenance
(ii) preventive maintenance?
(b) Give examples of the proper application of each, stating the advantages in all cases. (1969)

3. In order to improve the lubrication of plant in a factory, it is decided to employ a team of full-time operatives especially for the application of lubricants.
(a) What should be their duties?
(b) In what form should they receive their instructions?
(c) What information should they report back and how should they give this information? (1969)

4. (a) Choosing any type of factory and process with which you are familiar, draw a layout of the factory plant giving approximate dimensions.
Use this to assist in preparing an outline scheme of plant maintenance covering the following:
(i) lubrication for the factory
(ii) maintenance of the factory services.
Make any assumptions you feel to be necessary in order to clarify and justify your scheme.
(b) Give details of the maintenance operation involved in one specific item (e.g. compressor or conveyor) of plant referred to in (a). A specification must be given for the purpose of justifying your answer.
Describe briefly the most common failure associated with this item and indicate how its recurrence might be reduced. (1970)

5. Describe a planned maintenance scheme suitable for a factory of your choice. The following information should be given:
(a) the product or products of the factory
(b) all services provided by the plant engineer's department

(c) the workshop facilities available for the various plant engineering trades
(d) the number of production employees
(e) the number of employees in each of the different trades in the plant engineering department
(f) any special maintenance problems applying to this factory.

Mechanical, electrical and building trades should all be involved.

Your maintenance scheme should include typical completed forms for requisitioning maintenance, job specification, job instruction, job recording, summarizing the daily work of a tradesman, summarizing the work done in the department. Examples of inventory and schedule entries should also be given together with breakdown documentation.

Give a brief account of the methods which could be used for assessing the individual and overall performances achieved.

State the advantages accruing from the use of a planned maintenance scheme such as you have described. (1970)

6. Describe in some detail a planned maintenance scheme suitable for use in a factory employing 500 production workers and 75 shop floor maintenance workers. (1971)

7*. A new works to manufacture moulded rubber products comprises plant as indicated in the schedule below.

You are standing by during the plant installation and test, and will assume responsibility for the plant, excluding process machines.

The plant will operate on a daily two-shift system. It will shut down from Friday evenings to Monday mornings. It will also shut down for a week at Christmas and for a fortnight in the summer.

Set out in concise detail:

(a) the minimum requirements for operating and maintainence staff,
(b) daily operational routine, and records to be made and kept by yourself and staff (assume that the instruments which have been installed are adequate),
(c) daily inspection and maintenance routines,
(d) schedule of inspection and maintenance work at weekend shut-downs (observing staff requirements for reasonable leisure during the working week),
(e) schedule of major inspections and maintenance work during holiday period shut-downs,
(f) lists of stores and spare parts necessary to be kept for operation, maintenance and minor breakdowns (assume that for large overhauls and large breakdown repairs you may call on the makers or other specialist firms for assistance),
(g) fire precautions and initial action by the staff in the case of fire (assume adequacy of fire fighting equipment, fire alarms and communication with the local Fire Brigade).

Schedule of plant:

(i) A medium speed four-cylinder four-stroke single-acting compression-ignition engine provided with compressed air starter.

Engine cooling is provided by a circuit comprising motor driven pump, evaporative cooling tower with electric fan induced draught and thermo-

*Question originally set in Imperial units.

statically controlled by-pass valve. Make-up is from town mains, via a ball-float valve.

The engine drives a 250 kW dynamo supplying d.c. at 440 volts to a three-wire distribution system. The system is arranged to give 440 volts to motor driven plant and 220 volts to lighting and 13 amp power point sockets.

(ii) A vertical fire-tube waste-heat boiler (Cochran type) combining engine-exhaust heating and supplemented by oil-firing, fitted with the statutory mountings and additionally an auto-feed-regulator and whistle-type high-and-low water alarm.

The boiler generates saturated steam at 3·5 bar (50 lbf/in^2). The boiler is supplied by a motor-driven gear-pump with constant-pressure discharge by-pass valve taking water from a condensate return tank having a ball-float valve make-up supply from town mains.

Oil firing is of the air-atomizing automatic-type with motor-driven combined fuel-pump and air-blower, with steam-pressure control from output and electrically operated high-and-low water and flame-out cut-off to pump motor.

(iii) Externally-sited lagged tank for light fuel oil (1000 secs Redwood) for engine and boiler fitted with thermostatically-controlled steam-coils and electric immersion-heaters.

(iv) Steam and condensate lines.

(v) A two-cylinder, tandem two-stage air compressor of 1·4 m^3/min (50 ft^3/min) free air delivery at 6.2 bar (90 lbf/in^2) output with inter- and after-cooling, motor driven with integral circulating pump drawing water from and returning it to the cooling tower. This discharges, via oil, water and dirt separators to a 0·566 m^3 (20 ft^3) capacity air reservoir, from which the air passes through pipelines to the process plant.

(vi) Single-cylinder, tandem two-stage, air-cooled motor-driven air-compressor 0·14 m^3/min (5 ft^3/min) at 6·2 bar (90 lbf/in^2) output, discharging via the separators of the main compressor to the air reservoir.

(vii) Limited supply via rectifiers, of 440 volt d.c. from the Electricity Board for use in starting up and during shut-down periods. (1966)

8* A plant for the canning or deep freezing of vegetables and fruit is situated in a rural area and the plant listed below is the responsibility of the plant engineer for operation and maintenance.

The plant is in use during the months May to October inclusive, during which period it operates an eight-hour day on from three to seven days per week. During this period it is vital that breakdowns are at a minimum and should they occur be rectified with the minimum delay.

The plant is out of use over the period November to April inclusive, and during this period major inspection and maintenance is carried out.

Schedule of plant:

(a) One oil fired boiler of economic type (horizontal cylindrical with tubular furnaces, dry back combustion chamber and return fire tubes) with super heater.

The capacity of the boiler is 2270 kg/hr (5000 lb/hr) of steam at 12·75 bar (185 lbf/in^2), some of which is delivered saturated, and the remainder is super heated to 260 °C (500 °F).

Induced fan draught.

*Question originally set in Imperial units.

Mountings include auto-feed regulator, high and low level water alarm, steam pressure control of fuel and draught, and saturated and super heated steam stop valves.

Boiler ancillaries comprise fuel pump, heater and filter sets, motor driven feed pump, direct-acting steam feed pump, feed tank, fuel tank, valves and pipework.

(b) One two-cylinder compound reciprocating forced lubricated steam engine taking steam at 12·75 bar (185 lbf/in^2) and 260 °C (500 °F) driving a compound wound dynamo and exhausting to process line at 2·4 bar (35 lbf/in^2) dry saturated.

(c) (i) One compound wound dynamo supplying 100 kW at 440 V d.c.
(ii) One switch and distribution board supplying plant through three-wire system at 440 and 220 V.

(d) (i) One vapour compression refrigeration plant using CO_2 refrigerant, with water-cooled condenser and two evaporator circuits in parallel. One cooling brine circuit for the deep freezing plant. The other for cooling the cold store room.
The total capacity 50 tonnes (50 tons) of Refrigeration.
(ii) One fan draught evaporative cooling tower for cooling refrigerator-condenser water and cooling water for cans.

(e) Water supply comprising deep well pump, water tower, ionic treatment plant and pipes, valves, etc. to process, boiler feed, fire, domestic services, etc.

(f) Process steam supply comprising engine exhaust steam, boiler saturated steam reduced to 2·4 bar (35 lbf/in^2), valves to and steam traps from plant, reducing valve 2·4 bar (35) to 1·375 bar (20 lbf/in^2), valves and traps, flashpot and condensate return line through calorifier coil to feed tank. Calorifier and hot water storage tank.

(g) Battery starting compression-ignition engine driven stand-by dynamo 20 kW 440 V d.c. Water cooled, using cooling tower.

(h) Various centrifugal and gear type circulating and transfer pumps.
Lighting and power circuits.
Motors, starters and controllers which are not an integral part of process machines.

Make out in concise adequate details:

(A) A schedule for *each* of the following, during the summer operating period:
(i) the inspection and maintenance to be carried out regularly,
(ii) operating instructions for start-up, running and shutdown of plant.
Note: The cold store is empty at the end of each working day, all contents being transferred to a central depot.
(iii) Spare gear and stores necessary to ensure adequate maintenance and minimum 'outage' during this period.

(B) A schedule for *each* of the following during the winter shutdown period:
(i) the inspection and maintenance to be carried out to ensure plant is satisfactory for the next operational period,
(ii) spare gear and stores required to ensure no delays in repairs which may be found necessary.

(C) A scheme for keeping continuous records of running hours, inspection results, repairs and renewals for plant items.
Prepare a specimen item record card. (1967)

Bibliography

Those wishing to pursue the subject further may find the following publications useful.

1. BRITISH STANDARDS INSTITUTE (1964). *Glossary of General Terms used in Maintenance Organization.* BS 3811 : 1964.
2. DEPARTMENT OF HEALTH AND SOCIAL SECURITY (1969). *Hospital Technical Memorandum 12: Maintenance of Buildings, Plant and Equipment.* HMSO
3. DEPARTMENT OF HEALTH AND SOCIAL SECURITY – WELSH OFFICE (1967). *Hospital Technical Memorandum 13: A System for Engineering Plant and Services.* HMSO
4. FALCONER W. H. (1957) Some aspects of the application of planned maintenance to marine engineering. *Trans. Inst. mar. Engrs*, **69**, 2, 37
5. ROBERTS J. D. (1963). Steel works maintenance organization. *Proc. Instn mech Engrs*, **178**, 3g, 24
6. DOUGLAS-MORRIS K. J. and ELVIN R. H. P. (1964). The ship maintenance authority: a Naval approach to improving material availability. *Proc. Instn mech Engrs*, **179**, 3i, 142
7. WRIGHT V. J., SEYMOUR R. E. and NORRIS N. D. (1966). Maintenance planning applied to plant auxiliaries in the Central Electricity Generating Board. *Proc. Instn mech. Engrs*, **181**, 3n, 87
8. LUCAS A. H. S. *Planned Maintenance.* HMSO.
9. CORDER G. G. (1967). *Maintenance – Techniques and Outlook.*
10. STEWART H. V. M. (1963). *Guide to Efficient Maintenance Management.* Business Books Ltd., London.
11. SWARD K. (1966). *Machine Tool Maintenance.* Business Books Ltd., London.
12. MULHOLLAND J. R. (1970). *Heating, Ventilation and Air Conditioning Plant Planned Maintenance and Operation.* Business Books Ltd., London
13. CLEMENTS R. and PARKES D. (Ed.) (1966) *Manual of Maintenance. Vol. 1 Buildings and Building Services, Vol. 2 Plant.* Business Books Ltd., London
14. *Maintenance Engineering* – a monthly journal. Factory Publications Ltd., London
15. NIAZI A. A. (1968). *Management of Maintenance,* Asia Publishing House, London
16. CEGB (1971) *Modern Power Station Practice, Vol. 7 Operation and Efficiency.* Pergammon Press, Oxford
17. PARKES D. (1971). The problems of the maintenance engineer. *J. Instn mech. Engrs,* **18,** 5, 174.

Index